the ultimate

Baofeng Radio Bible

for beginners

An Easy-to-Follow Guerrilla's Guide to Master Your
Baofeng Radio and Protect Your Loved Ones

Joseph Maxwell

© Copyright 2024 By Joseph Maxwell- All rights reserved.

The content contained within this book may not be reproduced, duplicated, or transmitted without direct written permission from the author or the publisher.

Under no circumstances will any blame or legal responsibility be held against the publisher, or author, for any damages, reparation, or monetary loss due to the information contained within this book, either directly or indirectly.

Legal Notice:

This book is copyright protected. It is only for personal use. You cannot amend, distribute, sell, use, quote or paraphrase any part, or the content within this book, without the consent of the author or publisher.

Disclaimer Notice:

Please note the information contained within this document is for educational and entertainment purposes only. All effort has been executed to present accurate, up to date, reliable, complete information. No warranties of any kind are declared or implied. Readers acknowledge that the author is not engaging in the rendering of legal, financial, medical, or professional advice. The content within this book has been derived from various sources. Please consult a licensed professional before attempting any techniques outlined in this book.

By reading this document, the reader agrees that under no circumstances is the author responsible for any losses, direct or indirect, that are incurred as a result of the use of information contained within this document, including, but not limited to, errors, omissions, or inaccuracies.

ISBN: 9798329006575

An Easy-to-Follow Guerrilla's Guide to Master Your Baofeng Radio and Protect Your Loved Ones

the ultimate

Baofeng Radio Bible

for beginners

Joseph Maxwell

| Table of Contents |

Introduction   09

What This Little Radio Can Really Do   10

Which Radio Is Best for You?   12

The Four Most Valuable Radio Models
What's in the Box?
Keeping It Affordable
Future-Proofing & Special Use Cases
Myths & Mistakes to Avoid
Using Baofeng Internationally

Unboxing & Setting Up   21

Let's Build It: Step-by-Step Assembly
First Settings You'll Want to Check
Common Setup Mistakes to Avoid
Your First Radio Test
Storing, Charging, and Caring for Your Radio

The Easy Guide To Programming 29

How to Program Channels by Hand
How to Use a Programming Cable and CHIRP
What Channels Should
Common Programming Mistakes

Your First Transmission 37

Radio Etiquette for Beginners
What You Can Say (And What You Shouldn't)
When It's Time to Transmit &How to Do It Right
Common First-Time Mistakes

Getting Licensed Without Freaking Out 43

What the Technician Test Looks Like (It's Not That Bad)
A Simple 4-Day Study Plan for Busy People
The Only Study Tools You'll Need to Pass
How to Find and Schedule Your Exam
What to Expect After You Pass

Talking Like a Licensed Pro 50

How to Find People to Talk To
Making a Call: What to Say (And How to Say It)
Logging Contacts and Starting Your Radio Log
Handling Silence, Static, and Missed Calls

Master Your Radio's Hidden Tricks   59

Using Dual Watch and Dual Standby

Switching Transmit Power for Better Range or Battery Life

Improving Your Signal With Antenna Upgrades

Understanding CTCSS and DCS for Repeaters

Better Accessories

Cloning Channels With CHIRP (The Easy Way)

Using Your Radio in Real-Life Scenarios

Advanced Features for Everyday Use   68

Offset

How to Set Offset and Shift on Your Baofeng

Where Do You Find the Offset Info?

Priority Channel Monitoring

Wideband vs. Narrowband

Voice-Activated Transmission (VOX)

Smarter Scanning

Organizing and Labeling Channels

Hidden Extras That Might Actually Come in Handy

Finding Your Radio Rhythm   77

Appendix   87

Glossary   95

Introduction

Ham radio, also called amateur radio, has been around since the early 1900s. People all over the world have used it to chat, stay safe, and have fun. Today, over 2 million people are active on the air, some as a hobby, others as part of their emergency plans when the power or cell signal goes out.

You're probably holding this book because you want to learn how to use a Baofeng radio, and you're in the right place. Whether you're brand new or you've already dabbled in radio, this guide keeps things simple.

We're focusing on Baofeng models because they're affordable, tough, and incredibly popular, especially for beginners. In fact, plenty of folks stick with their Baofeng long after they've moved on to fancier gear.

You'll find everything you need right here, no endless Googling, no confusing tech talk. This book keeps things clear, practical, and beginner-friendly. And if something doesn't stick the first time, no worries. You can always come back for a quick refresh.

Don't have a radio yet? No worries, the illustrations inside will help you see what everything looks like while you're learning.

Some parts might feel confusing at first, and that's okay. Just keep going. This book repeats the really important stuff a few times to help it stick. Pretty soon, it'll all start to click.

You'll start by learning what Baofeng radios are and how to pick the right one. Then we'll get into unboxing it, putting it together, and setting it up. You'll learn how to use the buttons, tune into channels, understand the menus, and get ready for emergencies.

As we go further, we'll look at some of the cool features you can try, how to fix common issues, and how to take care of your radio so it lasts a long time. We'll also talk about using it in the real world, whether you're hiking, helping out during a storm, or just chatting with a group, and we'll point you to some handy extras and places to learn more.

Even though each chapter works on its own, if you're just starting out, it's a good idea to read everything in order the first time. That way, each step builds on the last, making it all easier to follow.

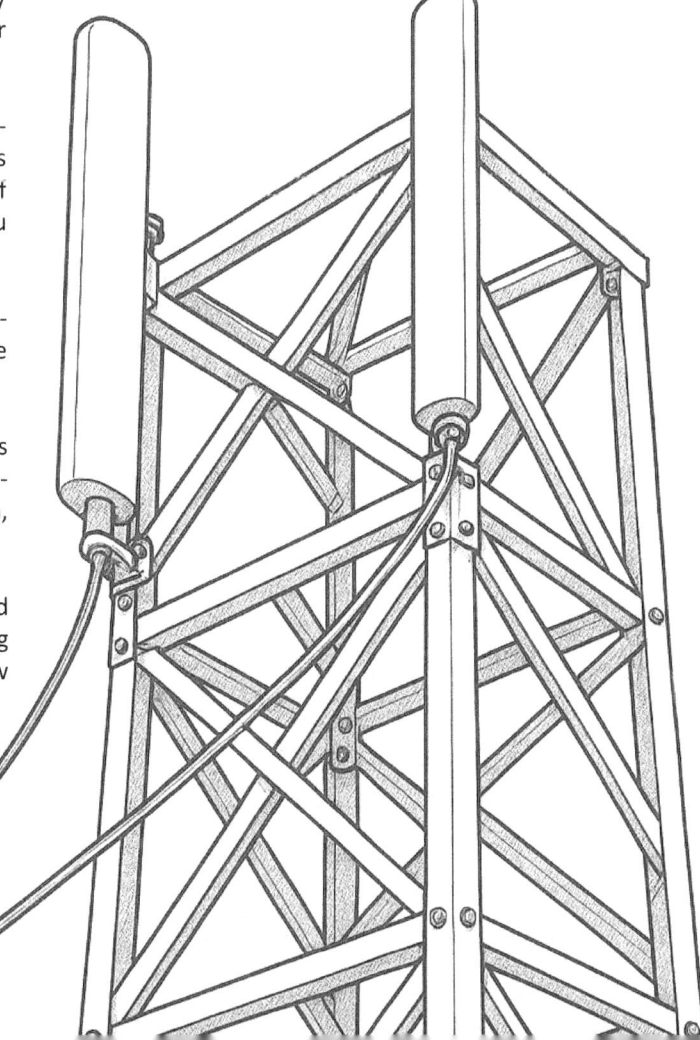

What This Little Radio Can Really Do

It might surprise you, but radios aren't old news, they're more important than ever. Sure, we all have phones in our pockets. But what happens when the cell towers go down? What if you're hiking in the mountains with zero signal? Or there's a blackout, and your only way to call for help is something that doesn't rely on the internet or Wi-Fi?

That's where ham radios come in. Also called amateur radios, these little devices let you talk directly to someone, even if there's no phone service, no data, no power grid. Just you, your radio, and a little bit of know-how.

And here's the cool part, anyone can learn to use one. You don't need to be a tech wizard. You don't need expensive gear. You just need a bit of curiosity and the right guide, and that's what this book is for.

People use ham radios for all kinds of reasons. Some join local radio clubs to chat with others around the world. Others rely on radios to stay safe during camping trips, road trips, or power outages. And plenty of people just enjoy the challenge, kind of like learning to drive stick shift or building your own campfire.

It's a skill, a backup plan, and honestly, it's fun. Now you're probably thinking, "Sounds cool, but I have no clue how any of this stuff works."

That's totally normal!

Let's start with just a few basics. No need to memorize anything, just get familiar with the words. We'll build from here.

A radio is just a device that lets you send and receive messages using invisible waves in the air. It's like a walkie-talkie, but with way more power and way more features.

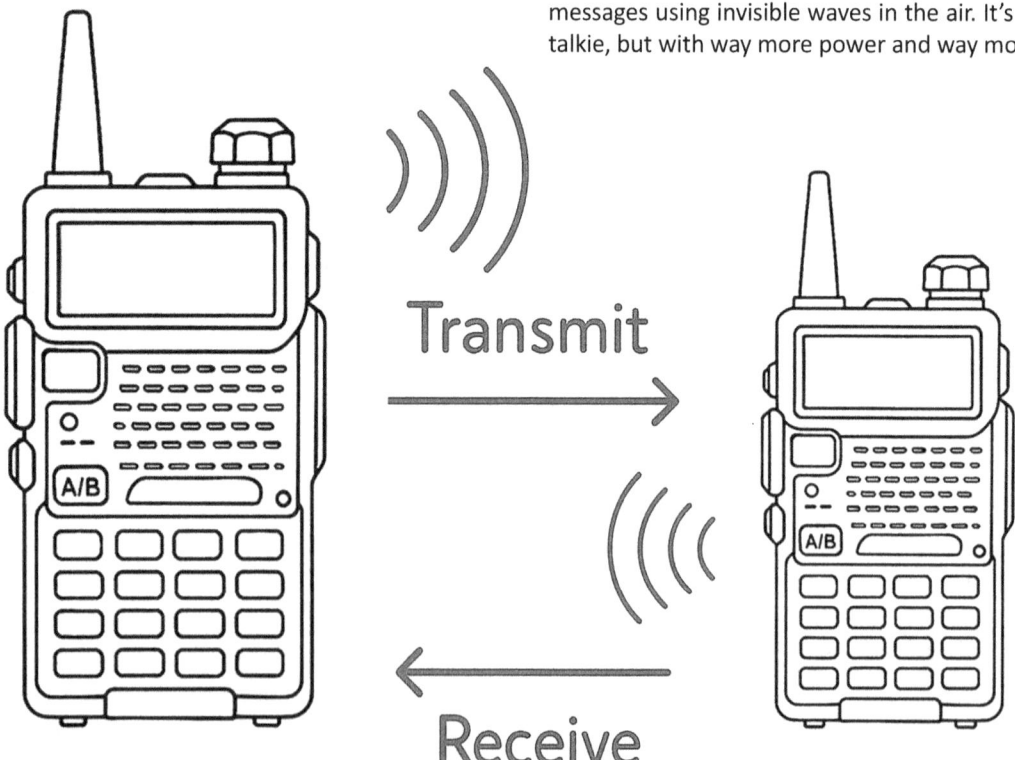

10

A channel is like a lane on a highway, and your Baofeng radio can hop between lots of channels(or lanes).

Now here's a funny word, squelch.

It's a setting that blocks out background noise when nobody's talking. Without squelch, your radio might sound like a washing machine full of bees. Set it too high, and you might miss weak signals. You'll learn how to adjust it so your radio only speaks up when it matters.

And then there's the antenna. That's the long stick on top. It's what lets your radio catch signals and throw them back out. A good antenna setup can be the difference between crackly noise and crystal-clear voices.

Now before we dive in and go step-by-step on using your radio, let's talk rules real quick.

In the U.S., ham radio is managed by the FCC, the Federal Communications Commission. They've got some ground rules to keep things running smoothly. Nothing scary, you're allowed to listen all you want. But to transmit on certain channels, like to talk to others, you may need a license.

We'll keep things beginner-friendly and point out what's safe and legal as we go.

As for picking the right Baofeng radio, don't worry, we'll cover that in Chapter 2. There are a bunch of models out there, but I'll help you figure out which one makes the most sense for your needs, budget, and goals.

So, what's it actually like to learn how to use one of these radios?

Here's the honest truth, at first, it might feel like your radio is speaking a secret language. It's got buttons, menus, beeps, and a voice that says weird things when you press stuff. You might look at the screen and wonder, "What the heck is that?"

That's totally normal!
Learning to use your Baofeng is kind of like riding a bike with gears for the first time. It feels awkward at first. Then one day, you shift a gear, pedal, and everything just clicks.

So don't worry if you're feeling a little unsure right now. You'll get this. All you need is a little curiosity, a little patience, and a radio that's turned on and ready to play.

In the next chapter we're diving into the fun stuff, how to choose the right Baofeng radio for you. There are lots of models out there, some simple, some loaded with features, and we're going to make it really easy to figure out which one fits your style, your budget, and what you actually want to use it for.

You're going to love that part!

Which Radio Is Best for You?

So, you're ready to get your first Baofeng radio. Nice move.

There are a lot of models out there, and yeah, some of them look super similar. That's where people get stuck. They scroll through product listings with names like "UV-5R," "BF-F8HP," or "UV-82," and think, wait, aren't these all the same thing?

Not exactly.

Let's slow things down and keep it simple.

Baofeng radios all do the same basic job, they help you talk to people. But different models are better for different situations. Some are great for emergencies. Some are perfect for hiking or road trips. Others come with a few more features if you want to grow into more advanced use later on.

Before we get into model names and features, take a minute to ask yourself a few questions.

Where do I plan to use this radio? Around town, on hikes, or during emergencies?

Do I want something simple to start with, or something that has room to grow with me?

What's my budget? Good news, Baofeng radios are affordable across the board.

The goal isn't to buy the fanciest radio, it's to find the one that fits you right now.

Baofeng makes a bunch of different radios. The good news is, most of them are pretty similar. The even better news is, you only need to look at a few of them to find your match.

Here's a simple breakdown of the most popular models, and what each one is best for.

UV-5R

The classic. The starter. The fan favorite.

Why people love it: It's affordable, simple, and there's a massive community of users who can help if you ever get stuck. There are tons of tutorials online, and it's usually the first model recommended in ham radio groups.

Best for: Beginners, casual use, and learning how radios work.

What to know: It's not the most powerful, and it's not waterproof, but it's more than enough to get started. If you're dipping your toe into radio, this is a safe bet.

BF-F8HP

The UV-5R's stronger sibling. Same shape, more muscle.

Why people love it: It gives you up to 8 watts of power, almost double what the UV-5R offers. That means you can reach farther, especially in open areas. Most bundles also include a bigger battery and a stronger antenna.

Best for: People who want more range or plan to use it outdoors regularly.

What to know: It costs a little more, but still won't break the bank. It's a great pick if you want stronger performance from day one.

UV-82

A little bigger, a little louder, a little tougher.

Why people love it: The UV-82 has two push-to-talk buttons, so you can talk on one channel and monitor another. It also has a chunkier build, which some people prefer—especially when wearing gloves.

Best for: Outdoor events, team activities, airsoft, or anyone who wants something more rugged than the UV-5R.

What to know: It's still easy to use, but it feels like a step up in build quality. Think of it as the "walkie-talkie" version of a Baofeng

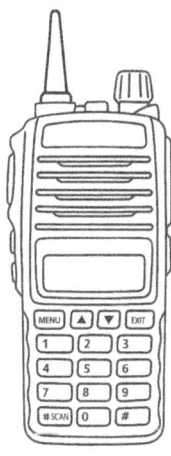

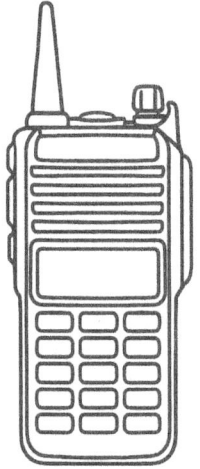

UV-9R / UV-9R Plus

The waterproof warrior.

Why people love it: Rain, dust, snow, this one's built to handle it. It's sealed tight and ready for messy conditions. It also has strong power output and solid battery life.

Best for: Hikers, campers, emergency kits, or anyone who wants to be ready when the weather turns bad.

What to know: Not all accessories fit this model, so double-check compatibility before buying extras. But if durability matters to you, it's hard to beat.

What's in the Box?

When you buy a Baofeng radio, you're not just buying the radio, you're buying a kit, and what's in that kit can vary depending on where you buy it and which bundle you choose.

That's why it's important to know what's usually included, and which accessories are worth paying attention to.

Let's unpack what you'll likely find when your radio arrives, and what each piece actually does.

1. The Radio Itself

This is your main unit, the part with the screen, keypad, and antenna connector. All models include it, of course, but it's still worth checking the model number stamped on the side to make sure you got what you ordered.

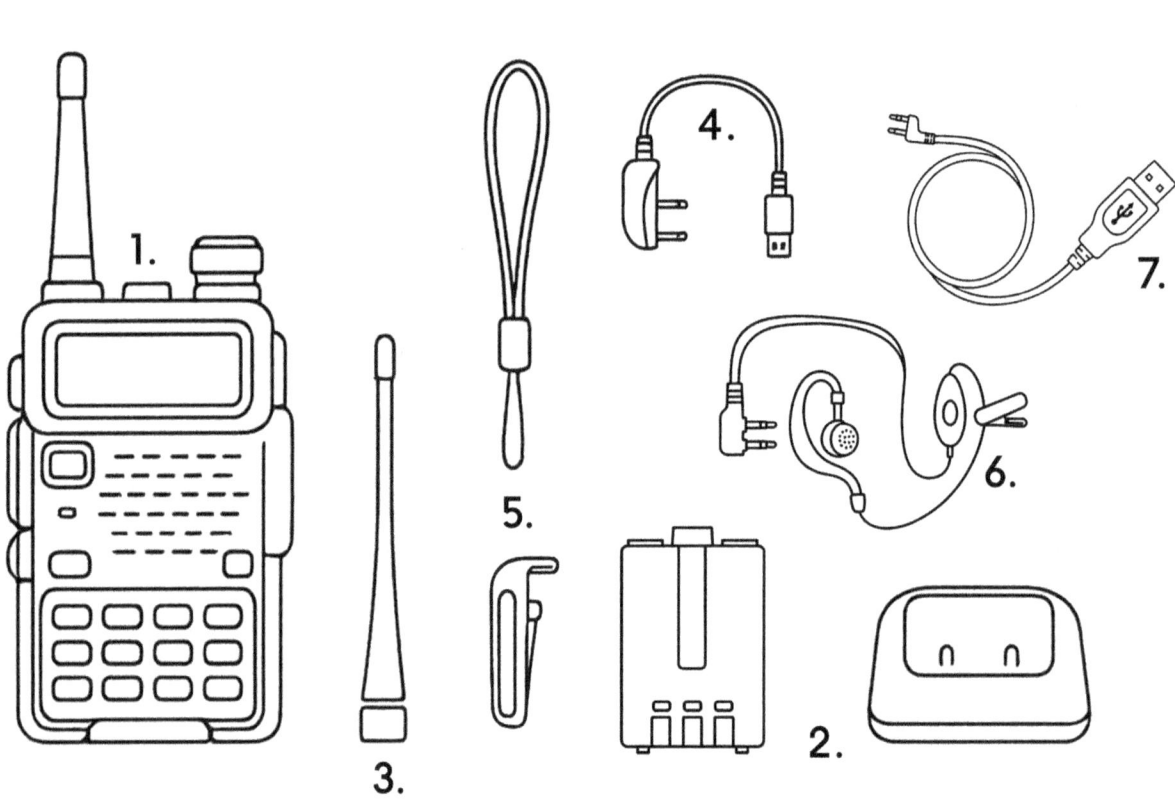

2. The Battery

Most Baofengs come with a rechargeable lithium-ion battery that snaps onto the back of the radio. Some bundles offer higher-capacity batteries, which are thicker and last longer. These are great if you plan to be out for long stretches without charging.

Tip: If you find a bundle that includes a spare battery, that's a major bonus, especially for emergencies.

3. The Antenna

All Baofeng radios include a basic rubber antenna, often called a "rubber duck." It works fine for most uses, but you'll get better range and clarity by upgrading later to a high-gain or whip antenna.

Pro tip: The BF-F8HP and UV-82 usually come with a better antenna right out of the box.

4. The Charger

You'll get a charging base, sometimes called a dock or cradle, and a wall adapter. Just drop the whole radio in, battery attached, and it charges up. Simple and tidy.

Watch for this: Some bundles include USB charging cables too, which are great if you want to charge from a power bank or in the car.

5. The Belt Clip and Wrist Strap

Small extras, but useful. The belt clip screws onto the back of the radio and lets you carry it on your hip. The wrist strap adds a little extra security if you're walking or hiking.

6. The Earpiece

Most kits include a basic push-to-talk earpiece, which is great for keeping conversations private or using the radio hands-free. Quality varies, so don't expect concert-grade audio, but they're useful for casual or team use.

7. Programming Cable (Sometimes)

This is the hidden gem of any good bundle. A programming cable connects your radio to your computer, so you can easily set frequencies and save channels using software like CHIRP.

Heads up: Not every kit includes one. If you find a deal that does, it's worth considering.

You don't need every accessory under the sun, but it helps to know what you're getting, and what you might want later.

If you're just starting out, a basic bundle with a battery, charger, antenna, and earpiece is usually more than enough. If you find a deal that includes a programming cable or an extra battery, even better.

Keeping It Affordable

One of the best things about Baofeng radios is that you don't have to spend hundreds of dollars to get something that works, and works well. That's a big reason why so many beginners choose Baofeng as their entry into the world of radios. It's affordable, and it gets the job done.

But just because something is cheap doesn't mean it's always a good deal.

If you're new to radios, you probably don't need the highest-watt, most feature-packed model right away. Look for Value, Not Just Price.

Sometimes the cheapest option isn't the best value. It's important to look at what's included in the box, not just the price tag. One listing might offer a UV-5R for $27 with no accessories. Another might cost $35 but include an earpiece, a high-capacity battery, a charger, a belt clip, and a programming cable. That second one gives you much more for your money and saves you the trouble of hunting down extras later.

When you're starting out, stick with analog radios. Digital radios offer features like better clarity and encryption, but they're more expensive and not always compatible with older equipment. Analog is simpler, easier to learn on, and perfect for nearly all amateur radio use.

You can also save money by focusing only on the accessories that matter at the beginning. You don't need a car mount, speaker mic, or tactical case right away. Start with the basics: a solid antenna (which you can upgrade later), a spare battery if you plan to be off-grid, and a programming cable to make setup easier on your computer. You can always add more gear once you know how you actually use your radio.

And if you see a deal that seems too good to be true, it probably is. Stick with trusted sellers. Plenty of knockoff models and off-brand clones look like real Baofengs but don't hold up. Always check for verified Baofeng listings, read customer reviews (especially ones that mention programming or actual use), and look for sellers that offer a warranty or return policy. A little caution up front saves you a lot of frustration later.

Future-Proofing & Special Use Cases

Buying your first Baofeng radio isn't just about today. If you pick the right one, it can be a tool you rely on for years, especially if you start exploring advanced features, join a radio club, or want to be better prepared for emergencies.

You don't need to max out your budget or grab the fanciest model on the shelf. You just need to make a smarter choice by thinking one step ahead.

One of the easiest ways to future-proof your radio is to pick a model that balances ease of use with room to grow. You want something that works great now but won't hold you back if you dive deeper later.

The BF-F8HP, for example, has a higher power output (up to 8 watts) compared to the UV-5R, giving you better range, especially outdoors or in hilly areas. It's still beginner-friendly, but you won't outgrow it quickly.

When you're choosing a radio that can grow with you, it helps to know which features really matter. Let's take a closer look at a few key things that make a big difference over time.

Power output: More power means more reach. It's not always necessary, but it's helpful if you move into more advanced use.

Dual-band capability: Most Baofengs already have this (VHF and UHF), which is perfect. Just avoid single-band models.

Rugged build: If you plan to use your radio outdoors or in emergencies, look for water resistance or a sturdier body, like the UV-9R Plus.

These features don't make a radio harder to use, they just make it more adaptable over time.

Some sellers offer radio kits with extras that go a long way. If you find a bundle that includes a programming cable, a high-gain antenna, a larger-capacity battery, or a desktop charger, you're getting more than just convenience, you're getting tools that help your radio grow with your needs.

Even better, many of these accessories are interchangeable between models, especially if you stay within the UV-5R family.

If you're buying with emergency preparedness in mind, a future-proof choice might mean a few extra features.

Weatherproofing, like what you get with the UV-9R, makes a huge difference for go-bags and outdoor use.
Longer battery life is another plus. Look for bundles that include spare or extended batteries.

Reliable signal power also matters. More wattage helps if you're reaching out over long distances when it counts most.

You might not need these features right away, but if you're building a kit for storms, power outages, or off-grid use, they'll pay off quickly.

You don't have to buy the biggest or most expensive radio on the market. But thinking ahead, even just a little, can save you from needing an upgrade too soon. Choose a model that checks today's boxes and leaves room for what's next.

Myths & Mistakes to Avoid

When you first start researching Baofeng radios, you're going to run into a lot of strong opinions, and not all of them are helpful. It's easy to get spooked by misinformation or feel like you're doing something wrong before you've even turned your radio on. So let's clear up a few common myths and beginner mistakes, right here and now.

One of the biggest myths out there is that Baofeng radios are illegal. They're not! What matters is how you use them. If you're in the U.S., you need a license to transmit on certain frequencies, like amateur radio bands. But just owning a Baofeng, turning it on, and even listening? Totally fine. In fact, many people use Baofeng radios as scanners to listen to local activity, weather updates, or emergency channels, all perfectly legal.

Another myth is that cheap means low quality. Yes, Baofeng radios are affordable, but that doesn't mean they're junk. They're tough, dependable, and wildly capable for the price. Thousands of hobbyists, preppers, outdoor enthusiasts, and even public safety volunteers rely on them every day.

Some folks will say Baofengs are "too complicated for beginners." Not true. They can seem a little overwhelming at first, especially if you're brand new to radio lingo, but that's why guides like this exist. With a little patience and a step-by-step approach, you'll be navigating menus, programming channels, and talking like a pro in no time.

Here's another common beginner mistake, skipping the user manual. Manuals aren't exciting, I get it. But Baofeng radios have quirks, and even a quick skim of the manual (or a good video walkthrough) will save you loads of frustration.

Another mistake is not getting a programming cable. Trust me, manually programming dozens of frequencies using tiny buttons isn't fun. A cable and free software like CHIRP can make setup a breeze.

The truth is, Baofeng radios are simple, effective, and built for learning. Don't let the noise online distract you. Focus on getting familiar with your radio, practicing a little each week, and learning at your own pace. You're not doing it wrong. You're just getting started, and that's exactly where you're supposed to be.

Using Baofeng Internationally

Baofeng radios are sold all over the world, but that doesn't mean you can use them the same way everywhere. If you plan to travel with your radio or use it outside the U.S., there are a few important things to know upfront.

First, radio frequencies are regulated country by country. What's perfectly legal to use in one place might be restricted, or even illegal, in another. That doesn't mean your Baofeng suddenly stops working when you cross a border, but it does mean you should do a little research before you power on.

The good news is, most Baofeng models cover a wide range of frequencies, often more than you'll ever actually need. That includes both the VHF and UHF bands, which are used worldwide for different types of communication. In many places, you'll be able to legally listen in or use certain frequencies as long as you follow local rules.

But transmitting without a license, or using frequencies reserved for emergency services, can still get you into trouble.

The best approach? Keep it simple.

Before using your radio in a new country, check the local regulations. Look up which frequencies are legal for amateur or general-purpose use, and avoid transmitting unless you're sure you're in the clear. Many countries have ham radio licensing programs similar to the U.S., and if you're licensed in one place, you might be able to operate under a reciprocal agreement elsewhere.

If you're just planning to listen, like tuning in to weather broadcasts or local radio traffic, you're probably fine. But if you're planning to use your Baofeng while traveling, camping abroad, or operating from a boat, it's worth double-checking that your setup won't cause problems.

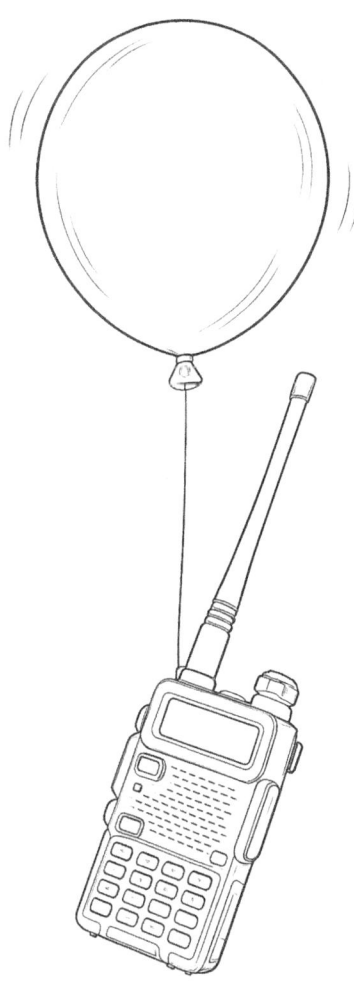

Chapter Quiz

How Much Do You Remember?

---

1. Which Baofeng model is considered the most beginner-friendly and affordable?

A) UV-82

B) UV-9R

C) UV-5R

D) BF-F8HP

2. What makes the BF-F8HP a better choice than the UV-5R for outdoor use?

A) It has a waterproof body

B) It comes in more colors

C) It has higher power output and better battery life

D) It's smaller and lighter

3. Which model features a dual push-to-talk (PTT) button, making it great for team communication and events?

A) UV-5R

B) BF-F8HP

C) UV-82

D) UV-9R

4. If you're planning to use your radio in wet, muddy, or harsh outdoor conditions, which Baofeng model is designed for that?

A) UV-9R / UV-9R Plus
B) UV-5R
C) UV-82
D) BF-F8HP

5. Why might you need to double-check accessory compatibility when choosing the UV-9R?

A) It doesn't use any accessories
B) It only charges via solar
C) It has a different connector than most Baofengs
D) It's not FCC-approved

6. The UV-5R is waterproof and ideal for rainy weather.
True / False

7. You need to spend over $100 to get a good Baofeng radio.
True / False

8. Some Baofeng radios let you listen to music on FM radio stations.
True / False

Answers: 1. C, 2. C, 3. C, 4. A, 5. C, 6. False, 7. False, 8. True

Unboxing & Setting Up

Now that you're ready to dive in, let's take a quick inventory. Open the box and lay everything out in front of you. It might look like a lot of little parts at first, but once you know what each one does, it all starts to make sense. Here's what you'll usually find:

The radio itself: The main unit with the buttons, screen, and antenna connector.

Battery pack: Usually comes detached to keep the packaging compact.

Antenna: A short, flexible rubber antenna that screws onto the top of the radio.

Charging dock and wall plug: Let you charge the radio by placing it upright in a small stand.

Belt clip and wrist strap: Optional but handy, especially if you're planning to take the radio on the go.

Earpiece with mic: Great for keeping things quiet or going hands-free.

Instruction manual: Hang onto this. Even though we're walking through it here, the manual has some useful reference info.

Some bundles also include extras like a USB programming cable, a high-capacity battery, or even a spare antenna. If you see those in your kit, bonus, you'll put them to good use later.

Lay the items out neatly, just like you would when setting up a new phone or camera. That way, nothing gets lost, and you can follow along step-by-step without scrambling to find a missing piece.

Let's Build It: Step-by-Step Assembly

Now let's put your radio together, one piece at a time. You're not installing anything complicated here, it's more like snapping Lego bricks into place.

Step 1: Attach the Battery

Pick up the battery pack and slide it into the back of the radio. You'll feel it lock into place with a soft click. If it wiggles or doesn't sit flush, remove it and try again. Don't force it, it should glide in smoothly.

Tip: Make sure the metal contacts on the battery align with the ones on the radio. That's how it gets power.

Step 2: Screw on the Antenna

Take the antenna and twist it onto the gold connector on top of the radio. It only needs to be snug, no need to crank it down super tight. A good rule of thumb: once it stops turning easily, it's tight enough.

Note: If you're using a longer or upgraded antenna (like a whip antenna), the same steps apply.

Step 3: Attach the Belt Clip (Optional)

If your kit includes a belt clip and screws, you can attach it to the back of the battery using a small Phillips screwdriver. It's handy if you'll be moving around a lot, but if you're just using the radio at home, feel free to skip it.

Step 4: Plug in the Charger

Take out the charging dock and plug the wall adapter into a standard outlet. Place the radio into the dock, battery still attached, antenna up. Most docks have a small indicator light to show charging status (usually red while charging, green when done).

If your radio came fully charged, the light may go green right away. But it's a good habit to start with a full charge before using it for the first time.

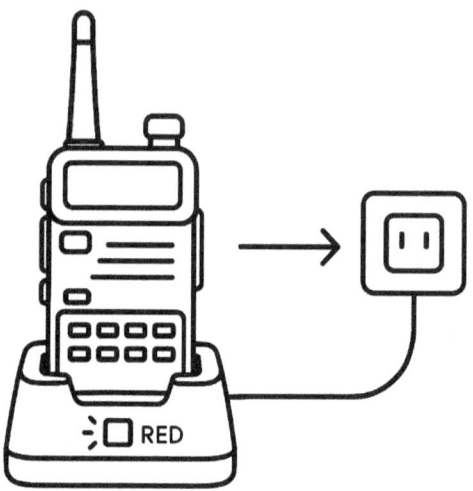

Tip: Always charge in a cool, dry place. Avoid windowsills or dashboards where it might overheat.

Step 5: Power It On

Once the battery is charged, press and hold the power knob. You'll hear a welcome beep and see the screen light up. That's it, your radio is alive.

Turn the knob slightly to adjust the volume, and you're ready to explore.

First Settings You'll Want to Check

Your Baofeng is now powered on, the screen is glowing, and the buttons are lit up. Nice work. Now let's make sure it sounds right and feels usable before diving into anything more advanced.

We're not digging into deep programming yet. This is just about getting your radio comfortable to use, like adjusting your car seat before hitting the road.

1. Adjust the Volume

At the top left of the radio, the same knob you used to power it on also controls the volume. Turn it gently until you hear a low hiss, that's background noise, and it means your speaker is working. Adjust it to a level that's clear but not too loud. You'll tweak it often depending on where you are, so just find a comfortable level for now.

2. Check the Squelch Setting

Squelch controls when the radio's speaker turns on. Too low, and you'll hear static all the time. Too high, and you might miss weak signals.

To adjust squelch:

- Press the MENU button.
- Use the arrow keys to scroll to SQL (usually Menu #0).
- Press MENU again to select it.
- Set it to around 2 or 3—a good middle ground.
- Press MENU to confirm, then EXIT to go back.

You can fine-tune this later, but a setting of 2 or 3 works great for general use.

3. Battery Check

Glance at the top corner of your screen, you'll see a small battery icon. If the radio came partially charged, it should show at least two or three bars. If it's nearly empty, return it to the charging dock for a bit longer before continuing.

Pro tip: If you ever plan to use your radio in emergencies, check the battery level regularly, just like you would check fuel in a generator.

4. Set the Display Mode (Optional)

Your screen may show frequency numbers or channel numbers (like "CH-001"). To switch between modes:

- Press VFO/MR to toggle between Frequency Mode and Channel Mode.
- Frequency Mode shows actual frequencies (like 146.520).
- Channel Mode shows pre-saved channels (like CH-005).

For now, it's fine to stay in Frequency Mode, you'll learn how to program channels soon.

5. Explore the Menus

You don't need to memorize anything yet. Just tap the MENU button and scroll through a few options. You'll see things like TXP (transmit power) and TOT (timeout timer). Don't change anything unless we cover it later, but getting familiar with the layout now will make learning easier.

With just a few adjustments, your radio is now ready for its first real use. You've turned it on, adjusted the volume, set the squelch, checked your battery, and even explored the menus a little.

You're not just holding a radio anymore, you're starting to understand it.

Common Setup Mistakes to Avoid

Even with something as beginner-friendly as a Baofeng radio, a few hiccups can sneak in during setup. The good news? They're all easy to fix, and even easier to avoid once you know what to watch for.

1. Not Clicking the Battery in Fully

Sometimes the battery feels like it's attached, but it's not actually connected. If your radio won't turn on, double-check that the battery has slid all the way into place and clicked. If it feels wobbly or loose, remove it and try again.

2. Over-Tightening the Antenna

The antenna only needs to be finger-tight. Cranking it down harder won't improve performance, and it can damage the threads or connector. Once it stops turning easily, you're done.

3. Forgetting to Charge Before First Use

Even if the battery looks full, it's smart to give it a solid charge before your first real use. A full charge helps the battery calibrate and perform better over time.

Make it a habit. If you're not using your radio daily, top off the battery every few weeks so it's ready when you need it.

4. Thinking It's Broken When It's Just on the Wrong Channel

It's surprisingly easy to end up on a quiet frequency and assume the radio isn't working. If you're not hearing anything, try scanning or manually changing channels. Most radios are set to random default frequencies that may not have any activity.

5. Ignoring the Manual (Then Getting Frustrated)

Manuals aren't thrilling, but Baofeng radios have quirks, and the manual can be surprisingly helpful. Keep it nearby, even if you prefer guides like this one. It's a good fallback when you hit a menu option you don't recognize.

Let's Try It Out: Your First Radio Test

Alright, your radio is set up, the battery's charged, the antenna's on, and you've adjusted a few settings. Now it's time to try it out for real.

Here's the good news: you don't need a license just to listen. Even if you're not ready to transmit yet, there's still plenty you can do right now to get comfortable and start learning how your Baofeng behaves.

Let's walk through a simple test anyone can do.

Step 1: Turn It On and Set the Volume

Press and hold the power knob until the screen lights up. Turn the same knob slightly to adjust the volume until you hear a soft hiss or background noise.

No hiss? You might need to lower your squelch setting. Remember, hearing a little static is actually a good sign, it means your speaker is ready.

Step 2: Scan for Activity

Tap the SCAN button. On some models, you might need to enter the menu first to activate scanning. Your radio will start scrolling through channels or frequencies, pausing whenever it hears a signal.

Stop and listen. You might catch ham operators chatting, weather reports, or even quick bursts of static from nearby equipment. Let it run for a few minutes to get a feel for how often something pops up.

If it stops on a channel with just noise, press the SCAN button again to keep it moving.

Step 3: Try Switching Channels

Press the up or down arrow buttons to scroll through channels manually.

- If you're in Frequency Mode, you'll see numbers like 146.520 or 462.5625.
- If you're in Channel Mode, you'll see labels like CH-001, CH-002, and so on.

Don't worry if you don't recognize the numbers yet. Just flip through and see if anything comes through.

Step 4: Press (But Don't Hold) the PTT

If you're curious about how the Push-To-Talk button works, go ahead and press it briefly, but don't transmit unless you're on an open channel you're legally allowed to use.

Just press and release. You'll hear a small beep, and the screen might flash. That's your radio switching into transmit mode.

Let go, and it returns to listening. Even this tiny action helps you start feeling how the radio reacts to touch.

Step 5: Explore the Dual Display

Most Baofengs show two lines of information on the screen. These are two separate channels you can monitor at once. Press the A/B button to switch between them.

This dual-display feature is great if you want to keep an ear on two channels at the same time, like monitoring weather and local chatter.

You won't need it right away, but it's good to know it's there.

That's It, You've Just Used Your Radio

You've turned it on, scanned for signals, explored channels, and tapped the talk button. That's a full test drive. You didn't need a license. You didn't need to program anything. But you're already ahead of most new radio owners, because you've actually used your gear.

Storing, Charging, and Caring for Your Radio

Now that your radio is up and running, let's make sure it stays that way. A Baofeng is a surprisingly tough little tool, but like anything else with a battery and buttons, it'll last longer if you show it a little care. You don't need to baby it, but a few smart habits will keep it ready whenever you need it.

Let's start with charging. When you drop your radio into the charging dock and plug it in, you'll notice a light, usually red while charging and green when it's full. As tempting as it is to leave it plugged in overnight, there's no need.

Once the light turns green, it's good to go. Leaving it on the charger too long won't instantly ruin the battery, but over time it can wear it down. Give it a full charge, then unplug it and let it rest.

Stick with the charger that came in the box if you can. Other chargers might fit, but not all of them work properly with Baofeng batteries. Using the original charger gives you one less thing to worry about.

One more thing, charge your radio somewhere safe and out of the way. A flat, dry surface is ideal. A windowsill or the dashboard of your car might seem convenient, but both can overheat your gear, especially in the summer.

When it comes to storing your radio, you've got a few easy options. If you're putting it in a drawer or backpack, keep the antenna attached. It protects the connector on top and saves time when you pull it back out.

Avoid extreme temperatures, which are rough on batteries. That means, no tossing it in a cold garage or leaving it to roast in your glove box. A closet, desk drawer, or shelf indoors works perfectly.

It's a good habit to leave the battery attached but keep the radio turned off. That way, it's ready at a moment's notice without slowly draining while it sits.

If you've got a pouch or small case, even better. It'll help keep dust out and give you a place to tuck in small extras like the earpiece or programming cable.

If you're keeping the radio in an emergency kit or go-bag, here's a simple routine to follow. Once a month, take it out, turn it on, check the battery level, run a quick scan, and give it a wipe-down if needed.

Five minutes is all it takes to make sure your radio is still in good working order, and that peace of mind is priceless when you actually need it.

If your radio ever gets wet, don't panic. If you've got a waterproof model like the UV-9R, it'll handle splashes and rain just fine.

If not, power it off right away, remove the battery, and dry everything thoroughly before turning it back on. The same goes for dust and dirt. A soft cloth and a quick wipe are all it takes. These radios are tough, but like any electronic gear, they'll last longer if you give them a little care now and then.

Chapter Quiz

How Much Do You Remember?

_____

1. What's the first step when assembling your Baofeng radio?

A) Scan for channels

B) Plug in the charger

C) Attach the battery

D) Screw on the antenna

2. What does a red light on the charging dock mean?

A) The radio is fully charged

B) The radio is off

C) The radio is still charging

D) There's an error

3. Where should you avoid placing your radio while it charges?

A) On a dry shelf

B) On your car dashboard or sunny windowsill

C) On a nightstand

D) In a drawer

4. What's the recommended squelch setting for general use?

A) 0-1

B) 4-5

C) 2–3

D) 8-9

5. If your radio turns on but you don't hear anything, what should you try first?

A) Change the battery

B) Lower the volume

C) Press the PTT button repeatedly

D) Check if you're on a quiet channel

6. Why should you keep the antenna attached, even when storing the radio?

A) It makes it look cooler

B) It helps protect the connector

C) It boosts Wi-Fi

D) It saves battery

7. What happens if you over-tighten the antenna?

A) You get a stronger signal

B) The antenna might snap

C) It could damage the connector

D) Nothing happens

9. You should leave your Baofeng plugged in 24/7 so it's always topped off.

True / False

10. The instruction manual isn't necessary once you've read this book.

True / False

11. You can switch between Frequency Mode and Channel Mode using the VFO/MR button.

True / False

12. Your Baofeng needs a ham license to be turned on and scanned.

True / False

_____

Answers: 1.C, 2.C, 3.C, 4.A, 5.C, 6.C, 7. False, 8. True

The Easy Guide To Programming

Let's take a deep breath and talk about a word that scares a lot of beginners: programming.

Don't worry, it's not as complicated as it sounds. You're not writing code, and you definitely don't need to be a tech wizard. Programming your Baofeng is more like saving your favorite stations in a car radio, simple, fast, and all about convenience.

It just means picking a few important channels you want easy access to, and storing them so you don't have to type them in every single time.

Remember earlier, when you flipped through random frequencies and scanned the airwaves? That was your radio in raw, live-listening mode, catching whatever it could find. But now imagine if you could lock in a few favorites: a local weather station, a nearby ham repeater, or your friend's walkie-talkie channel. Programming lets you do exactly that. It turns your Baofeng into a personalized hub tuned to what matters to you.

Here's the best part, once you save a few channels, everything gets faster and easier. Instead of memorizing a long number like 146.520, you can simply press the up or down arrow and jump straight to "Channel 1" or "Channel 2," already set and ready to go.

How to Program Channels by Hand (The Manual Way)

Let's walk through a real-world example. Say you want to save a basic frequency like 146.520, a popular national calling channel in the ham radio world. You'll store it so you can easily find it later as "Channel 1" or whatever slot you pick.

Step 1: Switch to Frequency Mode

Press the VFO/MR button to enter Frequency Mode. You'll know you're in the right place if the screen shows a full number like "146.520" instead of something like "CH-001."

Step 2: Enter the Frequency

Using the keypad, type in: 1-4-6-5-2-0. This tells the radio exactly what frequency you want to save.

Try This: Program in your local weather frequency (usually between 162.400–162.550 MHz) instead of a random one. To find yours go to www.weather.gov/nwr/station_listing and look up your city or region.

Step 3: Open the Memory Save Menu

Press the MENU button, then use the up or down arrows to scroll to MEM-CH (usually Menu #27 on the UV-5R, though it can vary slightly depending on your model).

The Easy Guide To Programming

Step 4: Pick a Channel Slot

Press MENU again to select it. You'll see a number like "CH-001" flash on the screen. If it shows "CH---" or an empty slot, it's ready for a new entry.

Step 5: Confirm and Save

Once you've picked a free channel slot, press MENU again to confirm. You'll hear a beep, and the screen might flash, that's your radio telling you it worked.

Step 6: Exit

Press the EXIT button to leave the menu and return to the main screen.

Step 7: Check Your Work

Press the VFO/MR button again to switch over to Channel Mode. Use the up and down arrows to scroll through your saved channels. If you find "CH-001" showing "146.520," congratulations, you've successfully programmed your first channel.

That's all there is to it. You've just manually programmed your very first frequency.

It might feel a little clunky the first few times, and that's completely normal. Everyone fumbles through the menu at first. It's part of the learning curve. But once you get the rhythm down, you'll be saving and organizing channels without even thinking about it.

Still, let's be honest, this button-by-button method isn't the fastest way to program a long list of channels. That's why many people prefer to use a simple programming cable and free software to handle it on a computer. And that's exactly what we're about to cover next.

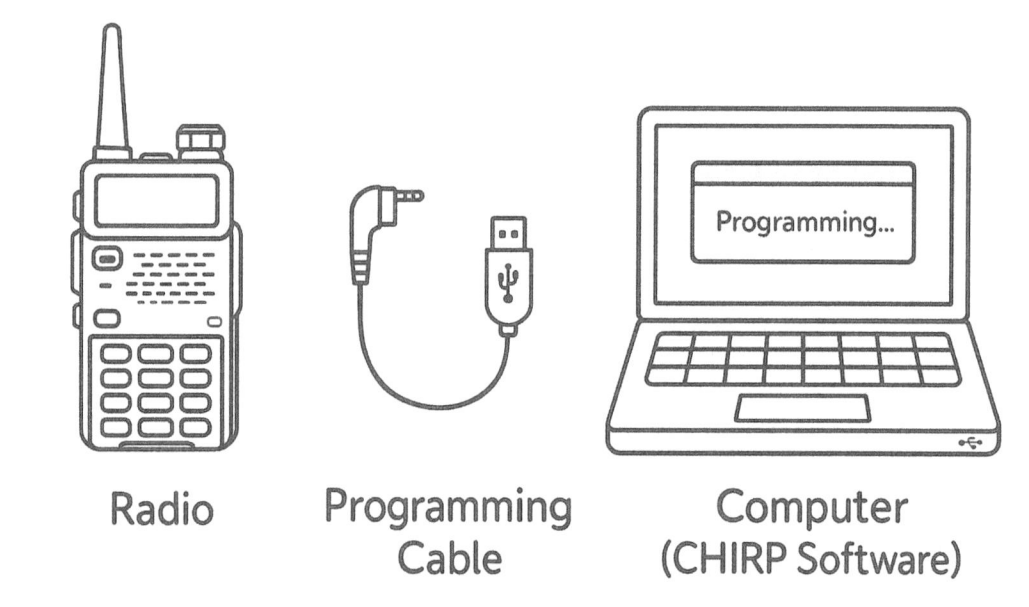

Radio   Programming Cable   Computer (CHIRP Software)

How to Use a Programming Cable and CHIRP (The Easy Way)

If manually programming your Baofeng felt a little clunky, here's some good news: there's a much easier way. With a simple programming cable and a free piece of software called CHIRP, you can load all your favorite channels onto your radio right from your computer. It's faster, easier to organize, and makes updating your setup a breeze later on.

CHIRP is one of the most popular tools in the radio world because it's beginner-friendly and works perfectly with most Baofeng models, especially the UV-5R and BF-F8HP. It's free, open-source, and trusted by thousands of radio users around the world. Think of it like a spreadsheet for your radio's memory: clear, editable, and backed up safely on your computer.

Ready to set it up?

Step 1: Get a Programming Cable

Look for a USB programming cable that matches your Baofeng model. One end plugs into the side of your radio (where the earpiece or mic normally connects), and the other end plugs into your computer.

Many beginner bundles already include this cable. If yours didn't, no problem, they're easy to find online. Just search for "Baofeng programming cable" and make sure it matches your model (most UV-5R and BF-F8HP models use the same type).

Step 2: Install CHIRP Software

Go to chirp.danplanet.com and download the latest version of CHIRP for your computer (whether it's Windows, Mac, or Linux). Follow the simple installation instructions.

Don't worry if the software looks basic, it's built for function, not fancy looks. That's part of what makes it fast and easy to use.

Step 3: Connect Your Radio to the Computer

Start by plugging the USB cable into your computer first. Then, connect the cable to your radio.

Important: Make sure the radio is turned off when you first connect it. Once the cable is securely in place, power the radio on. Your computer may take a few seconds to recognize the device, that's normal.

Step 4: Open CHIRP and Download From the Radio

Open CHIRP on your computer. Click "Radio" in the top menu, then select "Download From Radio."

A pop-up will ask you to choose your radio model (for most users, that's Baofeng → UV-5R or BF-F8HP) and pick the correct COM port.

Not sure which COM port to pick? No problem. Try each one in the list until it connects, CHIRP won't harm anything even if you choose the wrong one at first.

Tip: If CHIRP says "error" when connecting, it's almost always the COM port. Try the next one on the list, it won't hurt anything. Sometimes it takes a few tries, especially on laptops with multiple ports.

Step 5: Add or Edit Channels

Once your radio's memory loads, you'll see a list of channels laid out like a simple spreadsheet. You can click into each box to type in the frequency, add a name (like "Weather," "Local Rep," or "Hiking"), and set whether you want to transmit, receive, or both.

Want to expand your channel list even faster? You can import ready-made lists of local repeaters, weather stations, and emergency channels from online databases, saving you tons of time.

Step 6: Save and Upload

When you are happy with your channel list, click "Upload to Radio." CHIRP will replace your radio's current memory with the new setup you created. Before uploading, it is a smart idea to save a backup copy of your file, just in case you want to make changes or recover your old settings later.

Pro Tip: Always save your CHIRP file with a name like "Baofeng_StarterSetup_April25" so you'll remember when you last updated it. One file can save you hours if your radio ever resets.

Step 7: Unplug and Test

Turn off your radio, unplug the programming cable, and then power the radio back on. Use the up and down arrow buttons in Channel Mode to scroll through your saved channels.

If you see your new labels and frequencies showing up, congratulations, it worked perfectly.

Using CHIRP turns managing your radio into something as easy as setting up a playlist. It's simple to organize, quick to update later, and gives you a backup you can reload anytime if your radio ever resets or loses its settings.

What Channels Should You Save First?

Now that you know how to program your radio, the next step is deciding what to put in it.

You don't need a hundred channels to get started. In fact, adding just a few smart, useful ones will give you a strong foundation and help you feel confident that your radio is actually ready to use.

Here is a simple starter list to begin with. Think of these as your "favorites", frequencies that are active, practical, and perfect for beginners to monitor.

146.520 MHz, National Simplex Calling Frequency (VHF)

This is one of the most widely used ham radio channels in the United States. It is an excellent frequency to monitor, because many operators use it to call out and make contact before moving to another channel. Think of it as Channel 1 on a CB radio, a familiar place where people first reach out. Always a good place to start listening.

A Local Repeater (VHF or UHF)

Repeaters are large antennas, often placed on towers or hills, that boost the signal strength and greatly extend the range of handheld radios.

Nearly every area has a few public repeaters you can use. You can easily find local repeater listings online at websites like repeaterbook.com. Pick one close to your location and add it to your saved channel list.

NOAA Weather Radio Station (VHF, usually around 162.400 to 162.550 MHz)

NOAA weather stations broadcast around the clock and are essential for anyone interested in emergency preparedness, hiking, or outdoor events. These are receive-only channels, meaning you can listen without needing a license. Monitoring them keeps you updated on weather alerts and major changes in your area.

446.000 MHz, Common UHF Simplex Frequency

This is a widely used open frequency for short-range communication. You will often hear people using Baofeng or similar radios here, especially during outdoor events, field days, or local meetups. It is a good place to listen if you want to catch casual conversations without needing a repeater.

FRS/GMRS Channels (UHF, such as 462.5625 MHz or 462.7125 MHz)

These are the same channels used by most consumer-grade walkie-talkies. You can listen to these frequencies without a license, making them a great way to practice your listening skills. Keep in mind, transmitting requires a GMRS license or, in some cases, a very specific type of radio approved for those channels.

For now, feel free to monitor them safely and get a feel for the kind of local chatter you might hear.

Start by adding just three to five of these channels. Don't worry about filling all 128 memory slots your radio might support. A small, carefully chosen list is much better than a long scroll of static and silence.

After a few days of listening, you will naturally get a feel for which channels are active, and which ones you might want to replace or expand later.

If you are using CHIRP, you can label your channels with simple names like "CALLING," "WEATHER," or "LOCAL REP" to make them easy to spot at a glance.

If you are programming by hand, keeping a small notebook or printed list works just as well.

Common Programming Mistakes

Programming your Baofeng radio, whether by hand or with a computer, is a skill. And like any new skill, it's totally normal to fumble a few steps along the way. The good news? Most beginner mistakes are simple, easy to spot, and even easier to fix once you know what to look for.

Forgetting to Press MENU Twice When Saving a Channel

This is probably the number one beginner slip-up. You enter the frequency, scroll to a memory slot, and press MENU once, but unless you press it a second time to confirm, the radio won't actually save it.

Always press MENU once to open the save option, then press it again to lock it in. If you forget, no worries, you can always go back and try again.

Trying to Program in Channel Mode Instead of Frequency Mode

If your screen says "CH-001" or "CH-008," you're in Channel Mode, which won't let you type in new frequencies. Press the VFO/MR button to switch back to Frequency Mode first. Once you're there, you can start entering your numbers without a problem.

Saving Over an Old Channel by Accident
When you're programming manually, the radio won't warn you if you're about to overwrite something that's already saved.

That's why it helps to keep a small list of what you've programmed, and even better, to use CHIRP, where you can see your whole setup at a glance before making changes.

Using the Wrong COM Port in CHIRP

If CHIRP can't find your radio, don't panic. The problem is almost always the COM port setting. Just try selecting a different one from the list until you find the right match. You won't break anything by guessing, it's totally safe to experiment until the connection works.

Mixing Up Transmit (TX) and Receive (RX) Frequencies for Repeaters

When you're programming a repeater, you usually need two different frequencies: one to listen on, and one to transmit on.

If you enter the same number for both, your radio won't be able to reach the repeater. CHIRP makes this easy by handling the offsets for you, but if you're programming manually, double-check your TX and RX settings before saving.

Using a Knockoff or Poorly Made Programming Cable

Not all programming cables are created equal. Some cheap cables won't communicate properly with your computer, even if they look right on the outside. If CHIRP keeps throwing errors or refuses to connect, the cable is often the problem.

A good-quality cable, especially one with a known chip like FTDI, is worth spending a few extra bucks on, you'll save yourself a lot of frustration.

Not Saving a Backup Before Uploading in CHIRP

CHIRP gives you the option to save a backup of your current channel list, and you should always take it.

That way, if something gets erased, scrambled, or just doesn't turn out right, you can reload your old settings in seconds without losing all your work.

Mistakes aren't just common, they're part of the learning process. The key is to stay curious and not get discouraged when things don't work perfectly the first time. Your radio isn't broken, you didn't mess anything up, and there's almost always a simple fix waiting.

What's Next?

You've made it through what's probably the most technical part of learning your Baofeng, and you did it one step at a time.

By now, you understand what programming really means and, even better, why it matters. You've learned how to manually save a frequency to your radio, and you've seen how much easier it gets with a programming cable and CHIRP.

More importantly, you've started building a list of channels that actually matter to you. Whether it's a weather alert, a local repeater, or just a quiet frequency you want to monitor, your radio is no longer a blank slate. It's personalized. It's useful. It's yours.

Maybe you made a few mistakes along the way, and that's good. Every button press, every beep, every weird menu you backed out of in confusion, it's all part of the learning curve.

Now, you're no longer guessing what this radio can do. You're driving it. You've built a foundation that most new users never get, and you did it with patience, curiosity, and real hands-on practice.

Coming up next, we'll dive into the real heart of using your Baofeng: how to talk and listen like a licensed operator. You'll learn the rules, the lingo, and the simple techniques that help you sound clear, stay legal, and feel confident on the air.

Whether you're aiming for your ham license, getting ready for emergencies, or just curious about how real radio communication works, the next part will pull it all together. Your radio isn't just set up and programmed anymore, it's ready to connect.

Chapter Quiz

How Much Do You Remember?

_____

1. What does "programming" a Baofeng radio actually mean?

A) Changing the firmware
B) Installing apps
C) Saving favorite frequencies to memory
D) Entering Morse code

2. What mode must your radio be in before entering a new frequency manually?

A) Channel Mode
B) Monitor Mode
C) Frequency Mode
D) Scan Mode

3. When manually saving a frequency, what step do beginners often forget?

A) Entering the squelch setting
B) Holding down PTT while programming
C) Pressing the MENU button twice
D) Switching antennas

4. What does CHIRP do for your Baofeng radio?

A) Boosts the battery

B) Loads channels from your radio into your computer and lets you edit them

C) Acts as a voice translator

D) Blocks unauthorized transmissions

5. Which of the following is NOT a recommended "starter channel" to program?

A) Local weather station
B) A popular local repeater
C) 146.520 MHz (National Calling Frequency)
D) A random number you made up

6. What's one major benefit of using CHIRP over programming by hand?

A) It gives you free Wi-Fi
B) It uploads updates to your radio's firmware
C) You can see and organize all your channels like a spreadsheet
D) It works only on Android

7. CHIRP will permanently damage your radio if you choose the wrong COM port.

True / False

8. In Channel Mode, you can enter new frequencies using the keypad.

True / False

9. Saving a backup file in CHIRP is helpful in case your radio settings get erased.

True / False

10. You need to press MENU twice when manually saving a channel.

True / False

───────────────

Answers: 1. C, 2. C, 3. C, 4. B, 5. D, 6. C, 7. False, 8. False, 9. True, 10. True

Your First Transmission

You've already come a long way. Your radio's set up, charged, programmed, and ready. You've scanned through channels, maybe heard some voices, maybe just a lot of static. Either way, you've stepped into the world of radio, and now it's time to talk about, well, talking.

Here's the secret, using a Baofeng to talk is surprisingly simple. If you've ever used a walkie-talkie, you already know the basic rhythm. Press the button, talk, let go, listen.

But here's the difference, you're joining a space where people are careful, respectful, and purposeful about what they say. That might feel intimidating at first, but it doesn't have to be.

The most important thing to remember is this: you don't have to talk right away. In fact, it's better if you don't. Start by listening. That's how you learn what a real conversation sounds like.

That's how you pick up the unspoken rules, who speaks when, what they say, how long they take, and how they sign off. You'll get used to the pace and the tone just by paying attention. When you do feel ready to speak, it's easy to get started.

Step 1: Turn on your radio and pick an active channel. Choose a repeater or a calling frequency like 146.520.

Step 2: Wait and listen. Give it a few seconds of silence to make sure the channel's clear.

Step 3: Press and hold the Push-To-Talk (PTT) button. Speak clearly into the mic. Slow and steady is best, there's no rush.

Step 4: Release the button to listen. Don't forget to let go, if you keep holding the button down, no one else can reply.

Step 5: Repeat as needed. Keep your messages short, clear, and to the point. Always leave a few seconds between transmissions in case someone else wants to jump in.

And that's it. You've just done what every radio operator does, talk and listen. Simple, right?

Radio Etiquette for Beginners

Once you start listening to real radio conversations, you'll notice something right away, everyone has a rhythm. There's a certain way people speak, a calm tone they use, and a polite back-and-forth that keeps everything running smoothly. That's not just style, it's etiquette.

Think of radio etiquette as your on-air manners. It's not about being formal or stiff. It's about knowing when to talk, how to talk, and how to leave space for others. And the best part? It's all very easy to learn. Here are the basics to follow whenever you get on the air:

Always listen before you speak. Give the channel a few seconds of silence before you key up. Someone might already be in the middle of a conversation, and you don't want to interrupt.

Keep your messages short and clear. Radio isn't like a phone call. Stick to a sentence or two. Say what you need to say, then let go of the button and listen.

Speak slowly and clearly. When you press the PTT button, wait a half-second before talking. That helps make sure the beginning of your message doesn't get cut off. Talk at a natural pace, not too fast, not too soft.

Pause between transmissions. After someone finishes talking, wait a second or two before replying. That gives others a chance to jump in or call out if they need to.

Use plain language. You don't need to say "over and out" or use movie-style lingo. Just talk like a normal person. Clear, simple words are best.

Identify yourself. If you're licensed, use your call sign. If not, and you're just practicing, use your name or a placeholder. For example: "This is Alex, listening on 146.520."

Don't hog the mic. Think of it like a group conversation, you're taking a turn, not holding a lecture. Leave space for others and stay polite, even if nobody else is talking.

These small habits go a long way. They show respect for the people already using the airwaves, and they make you someone others actually want to talk to.

And if you make a mistake? That's okay. Most radio users are friendly and happy to help newcomers, especially those who are clearly trying to do it right.

You don't have to be perfect. You just have to be thoughtful. Keep your transmissions short, your ears open, and your voice steady. That's all it takes to sound like you belong.

What You Can Say (And What You Shouldn't)

Now that you've got the hang of listening, and you understand the flow of a typical conversation, it's natural to wonder, what exactly are you supposed to say when you key up the mic?

The good news is, you don't need a script. You just need a sense of purpose and a little respect for the space you're speaking into. Radio channels are shared, public airwaves. That means anyone could be listening, including experienced operators, beginners like you, or folks just passing through.

So the goal isn't to sound perfect, it's to sound clear, polite, and intentional.

Start with something simple. If you're making a general call on a repeater or simplex frequency, you could say:

"This is Alex, monitoring 146.520. Anyone around?"

That's it. You've identified yourself, mentioned the frequency, and let others know you're open to a chat. If someone responds, just keep it casual and clear. Say hello, maybe mention you're new, and let the conversation flow naturally. Most people will be friendly, especially if they know you're learning.

Now, as for what not to say, think of it like chatting at a coffee shop. No yelling. No swearing. No long personal monologues. Keep things appropriate and helpful.

This isn't the place for background music, random rants, or inside jokes. And while you might hear some of that from others now and then, don't follow their lead, especially if you're trying to build good habits from the start.

If you're not licensed yet, you should avoid transmitting on ham frequencies. But you can still listen, practice your message with the mic unpressed, or even role-play a call with a friend on an open channel. Every bit of practice helps.

In short, stick to clear, friendly communication. Identify yourself, keep it short, and speak like someone who deserves an invitation back.

Once you know what to say, and what to avoid, you're ready to actually make your first real transmission.

When It's Time to Transmit & How to Do It Right

So you've listened, you've learned the rhythm, and you've figured out what to say. Now it's time to actually press the button and talk. It might seem like a small moment, but for a lot of beginners, it feels like a big deal. The truth is, your first transmission will feel a whole lot less intimidating once you know exactly what to expect.

Here's a simple, beginner-friendly guide to making your very first call, whether you're reaching out on a repeater or trying a calling frequency.

Step 1: Pick a channel. Choose a frequency you know is active but currently quiet. If you've already programmed in a local repeater or a calling frequency like 146.520, that's the perfect place to start.

Step 2: Listen for at least 10 seconds. Make sure the channel isn't already in use. Give it a little breathing room, just in case someone else is about to speak. Waiting a few seconds shows good manners, and might save you from accidentally cutting someone off.

Step 3: Press the Push-To-Talk (PTT) button and hold it. Once you're ready, press and hold the Push-To-Talk button. Wait for about half a second before speaking. That quick pause gives your radio time to start transmitting, so your first words don't get chopped off.

Step 4: Speak slowly and clearly. Try saying something like: "This is Jordan, monitoring 146.520. Anyone available for a quick radio check?"

It's simple, it's clear, and it invites a friendly response without putting pressure on anyone to jump in.

Step 5: Release the button and listen. After you finish speaking, let go of the Push-To-Talk button and listen. Wait a few seconds. If someone responds, you might hear something like, "Copy that, Jordan, you're coming in loud and clear." From there, just reply in the same calm, clear tone.

Step 6: If no one responds, don't worry. You didn't do anything wrong. Sometimes the airwaves are just quiet. Try again later, or switch to another frequency. Even experienced operators get silence now and then, it's totally normal.

The first few times you do this, your heart might race a little. That's okay. You're stepping into something new, and you're doing it right.

Common First-Time Mistakes

If you've made it this far, first of all, well done. You've learned how to talk, how to listen, how to make a basic call, and how to treat the airwaves with respect. That's more than most people ever do with their radios. But let's be real, when you start transmitting, a few little slip-ups are bound to happen.

That's not failure, that's learning.

You might press the PTT button and start talking too quickly, cutting off your first word. You might forget to identify yourself, or accidentally say too much at once. Maybe you'll speak too quietly, too fast, or forget to leave space for someone else to jump in. Maybe you'll pause a little too long and miss your turn entirely.

Or maybe, and this is pretty common, you'll freeze up when someone actually answers you. You weren't expecting it, and suddenly you're not sure what to say.

All of that is totally okay. The people who use radios regularly? They've all been there. Most will be patient, especially if you're polite, trying your best, and willing to learn.

And if you make a mistake and realize it, you can always say something like, "Sorry, still getting the hang of this," and people will understand.

The important thing is that you don't let those little stumbles stop you. You're doing something new. You're learning a skill that takes time, patience, and a little practice. And every time you pick up the radio, listen, speak, or even just think about what you'd say next, you're getting better.

So go ahead, make mistakes. Learn from them.
Just keep going.

Chapter Quiz

How Much Do You Remember?

_____

1. What's the first thing you should do before talking on any frequency?

A) Press the PTT button
B) Start scanning
C) Listen for a few seconds
D) Adjust the volume

2. What's a good example of an opening message on a calling frequency?

A) "Hello? Hello? Anyone?"
B) "Is this thing on?"
C) "This is Alex, monitoring 146.520. Anyone around?"
D) "Testing 1 2 3, testing 1 2 3."

3. What should you do immediately after you finish talking on the radio?

A) Power off the radio
B) Press the MENU button
C) Change frequencies
D) Release the PTT and listen

4. What's a common mistake new users make when pressing the PTT button?

A) Holding it too long
B) Not pressing it at all
C) Speaking too soon and getting cut off
D) Forgetting where the button is

5. Which of the following should you NOT do during a radio conversation?

A) Keep your messages short
B) Use plain language
C) Yell or rant
D) Leave space for others to talk

6. If someone responds to your radio call and you freeze up, what should you do?

A) Turn off your radio
B) Apologize and try again
C) Ignore them
D) Say a random phrase quickly

7. You should hold the PTT button and start talking immediately.
True / False

8. It's okay to listen for a while before ever talking on the radio.
True / False

9. Saying something like "This is Jordan, listening on 146.520" is a polite way to identify yourself.
True / False

Answers: 1. C, 2. C, 3. D, 4. C, 5. C, 6. B, 7. False, 8. True, 9. True, 10. False

Getting Licensed Without Freaking Out

You've made it through the setup, the programming, the listening, and maybe even your first few transmissions. By now, you've gotten a taste of what your Baofeng can do, and you've probably noticed a few places where you're still holding back.

Maybe you've heard people talking on repeaters and wondered if you could jump in. Maybe you've scanned through channels and thought, I'd love to talk here, but can I? Or maybe you've just realized how powerful your radio really is, and you want to be sure you're using it the right way.

That's where getting licensed comes in.

Getting your amateur radio license, specifically the Technician Class license here in the U.S., isn't about becoming a hardcore hobbyist or memorizing obscure rules. It's about giving yourself permission to use everything your radio was built for: legally, confidently, and without second-guessing whether you're crossing a line. With a license, you can:

- Transmit on ham bands your Baofeng already supports

- Use local repeaters to reach farther than your handheld can on its own

- Join emergency comms groups, ham nets, and local clubs

- Talk with operators in your region, or even across the country

- Explore new parts of the radio hobby you haven't even touched yet

And best of all? You get to be part of a global community of people who love helping others, learning new things, and staying connected, especially when phones or the internet go down.

What the Technician Test Looks Like (It's Not That Bad)

Let's talk about the actual test, the thing standing between you and your ham radio license. You don't need a science degree or a lifelong love of electronics. If you've made it this far in this book, you've already done the hardest part, getting familiar with how radios work in real life. Here's what the test looks like:

- The test has 35 multiple-choice questions. You only need to get 26 right to pass, that's about 75%, or a solid C.

- Every question gives you four possible answers, and you just pick the correct one.

- There's no Morse code, no essays, and no hands-on practical exam. It's just a simple quiz.

- You can take the test either in person with a local group, or online from home, depending on what's available near you.

- It usually costs between $15 and $35, and once you pass, your license is good for 10 years.

The questions come from a fixed pool, about 400 possible questions in total. So when you use a practice app or take an online test, you're actually working with the exact same questions that might show up on the real thing. There are no surprises. If you practice a little each day, you'll start seeing the same questions again and again, and that repetition makes the real test feel easy.

Most of the topics fall into a few simple categories:

- Basic radio terms and how they work

- Safety and interference

- Rules about frequency use and who can talk where

- Some super light technical info (think: what an antenna does or what a repeater is)

And here's a bonus: **you don't even need to memorize every answer.** You just need to get familiar with how the questions are worded and learn to spot the right choice. Most practice apps show you the correct answer immediately, so every test becomes a lesson.

A Simple 4-Day Study Plan for Busy People

You don't need weeks of studying to pass the Technician exam. You just need a few focused sessions where you learn the basics, get familiar with the questions, and build up the confidence to hit "start" on test day.

Here's a simple four-day plan that works, whether you're squeezing it in after work, during your lunch break, or over a quiet weekend. You can stretch it out if you need more time, or double up if you're in a groove.

Day 1: Get Familiar With the Basics

Start by reading through a free study guide or watching a quick intro video about what the test covers. You're not trying to memorize anything yet, just getting a feel for the topics. You'll see things like frequency bands, what repeaters do, and how to avoid interference. Let it wash over you.

If you prefer reading, check out the KB6NU "No Nonsense Technician Study Guide." It's clear, short, and free. If you like learning by video, search YouTube for "Technician license explained for beginners." Pick whichever style fits your brain best.

Day 2: Take a Practice Test (and Don't Stress About the Score)

Now it's time to jump into a full practice test. You can do this online at sites like hamstudy.org, or by using a mobile app like Ham Radio Prep or HamTestOnline.

Don't try to get a perfect score right away. Just take the test and see what comes up. Every question you miss teaches you something valuable. Afterward, go through the missed answers, read the explanations, and move on. You'll already start spotting patterns without even trying.

Day 3: Review and Retake the Practice Test

Take another practice test today, but this time, slow down a little when you hit a question you've seen before. Can you remember the answer without guessing? If not, take a second to reread the explanation and really lock it in.

Then finish the test and check your score. You'll probably notice some improvement already..

Day 4: Final Review and Schedule Your Exam

Take one last practice test today, and pay close attention to your score. If you're consistently hitting 80% or higher, you're ready to go.

Now's the time to schedule your exam. I'll walk you through that in just a bit, but trust me, locking in a date keeps your momentum strong.

If you're not quite there yet, no worries. Keep taking practice tests, focus on the tricky spots, and move at your own pace.

The Only Study Tools You'll Need to Pass

Now here's the best part...you don't need to buy a big textbook or sign up for some fancy course to pass the Technician exam. There are tons of free (and low-cost) tools out there that make studying simple, and even kind of fun.

I've tested a bunch of them myself, and here are the ones I think work best for beginners. You only need one or two to succeed, so pick whichever feels easiest and dive in.

1. HamStudy.org (Free, Web-Based)

This is one of the best online study tools out there. It lets you take full practice tests, focus on specific types of questions, and track your progress over time. It even highlights the questions you keep missing, so you know exactly where to focus.

Just head over to www.hamstudy.org, select the Technician test, and either study by topic or dive straight into practice tests.

2. KB6NU's "No-Nonsense Technician Study Guide" (Free PDF)

If you like reading and want something short, simple, and clear, this is the guide for you. It explains everything in plain English without any extra fluff. You can finish it in just a couple of hours and come away thinking, "Oh, this really isn't that hard."

Just search "KB6NU Technician Guide" online and download the free version straight from his website.

3. Ham Radio Prep App (Free + Paid Options)

If you prefer studying on your phone, this app is a great option. It includes short video lessons, flashcards, and tons of practice questions. You can start with the free version, and upgrade later if you want a full step-by-step course.

It's available for both iOS and Android, just search for "Ham Radio Prep" in your app store.

4. ARRL License Manual (Paid, Print)

If you like studying with a physical book and want to dig deeper, the ARRL Technician License Manual is the official prep guide. It's more detailed than most beginners need, but perfect if you're looking for a full deep dive into the material. You can find it on Amazon or order it directly from the ARRL website.

5. Local Clubs and In-Person Classes (Often Free)

A lot of ham radio clubs offer free in-person or online classes, and some even host study groups to help you get ready for exam day. If you prefer learning with others, it's definitely worth searching for local clubs in your area to see what they offer.

You can find clubs near you by visiting arrl.org/find-a-club.

Once you've found a study tool that fits your style and started practicing a little each day, the next step is simple: get yourself on the calendar. Scheduling your exam might sound like a big deal, but it's easier than you think, and it's one of the best ways to stay motivated and lock in your momentum.

How to Find and Schedule Your Exam

You've studied, taken a few practice tests, and now you're ready to go. The final step? Actually signing up for your Technician exam.

The good news is, you've got plenty of options. You can take the test in person through a local ham radio club, or take it online from home, right at your desk. Either way, the process is simple, and once you're done, you'll have your license faster than you might expect.

Step 1: Get an FCC Registration Number (FRN)

Before you can take the test, you'll need to register with the FCC and get an FRN. Think of it as your ID number for anything radio-related.

- Just go to the FCC's CORES site: https://apps.fcc.gov/cores/userLogin.do

- Click "Register" and fill out your basic info.

- You'll get a unique 10-digit FRN, be sure to save it, you'll need it when you sign up for your exam.

Step 2: Choose How You Want to Take the Test

In-person tests are run by local Volunteer Examiner teams, usually through ham radio clubs. They're a great option if you prefer face-to-face testing and want to meet other radio folks.

Online tests are also available through groups like GLAARG, HamStudy, or ARRL VEC. These are proctored by video call, and you'll just need a webcam and a quiet space to take the test.

Step 3: Find a Session That Fits Your Schedule

Head over to hamstudy.org/sessions or arrl.org/exam_sessions and browse through the upcoming exam dates.

When you find one that works for you:

- Register using your name and FRN

- Pay the exam fee (usually between $15 and $35)

- Watch your email for a confirmation and any instructions

If you're testing online, they'll walk you through how to prepare your space, usually just a clean desk and a working webcam. For in-person tests, simply show up with your ID and confirmation email.

Step 4: Take the Test

Show up on time, stay calm, and take your time answering each question. You've already seen most of them while practicing, so nothing should feel like a surprise. It usually takes about 20 to 30 minutes, and many people finish even faster.

Once you pass, your results will be submitted to the FCC. Within a few days, you'll have your callsign assigned. You'll be officially licensed, and ready to transmit like a pro.

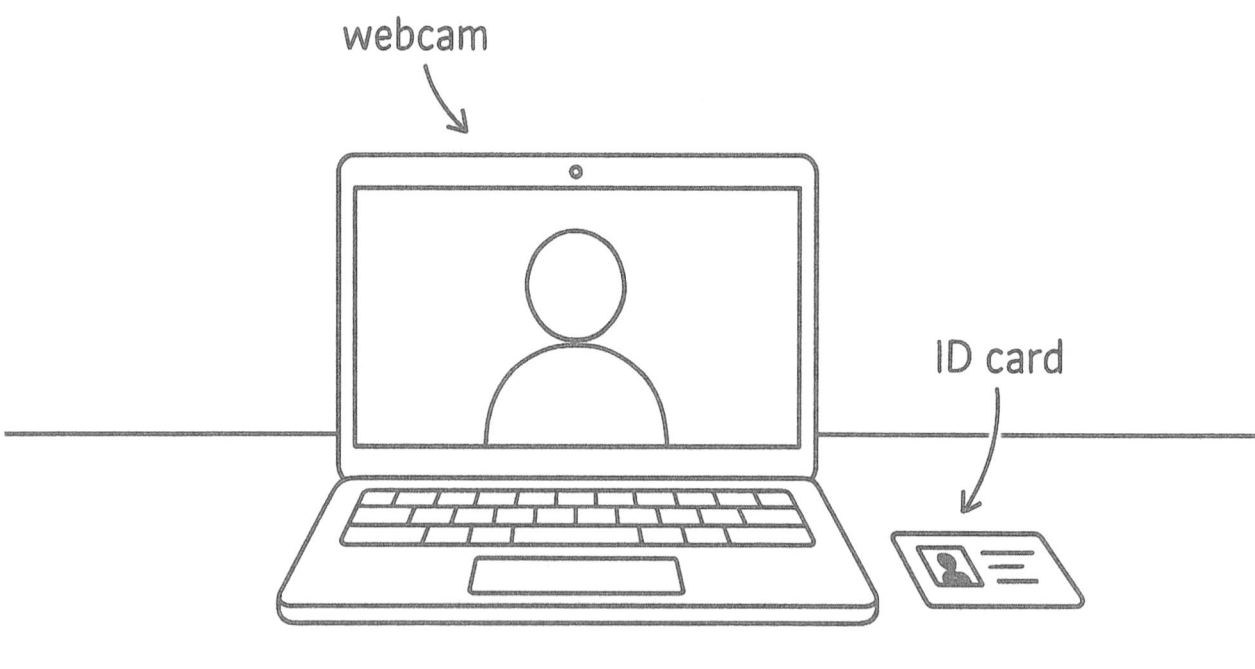

What to Expect After You Pass

You've taken the test, answered that last question, and heard the best news, congrats, you passed!

First, take a moment to feel proud. This is more than just passing an exam. You've learned a new skill, stuck with it, and earned the right to use your radio legally and confidently. You're not just a Baofeng user anymore, you're a licensed radio operator.

After you pass, your Volunteer Examiners will send your results to the FCC. If you tested in person, it usually takes a few business days. If you tested online, it might be even faster. You'll get an email once your license becomes active.

You can also look up your callsign yourself. Just head to the FCC's license search site and enter your name or FRN. When your name pops up, you'll see your new callsign right next to it. That's your official on-air identity. Once your callsign is issued, here's what you can do right away:

- Start transmitting on all the bands available to Technician licensees, no more holding back
- Legally access and talk on repeaters
- Check in to ham nets (on-air meetups that happen daily or weekly)
- Join a local ham radio club and participate in events, field days, or emergency training
- Help your community during weather events or outages
- Upgrade your gear, or stick with your trusty Baofeng, now fully unlocked

There's something powerful about saying your callsign on-air for the first time. It's your entry into a tradition that's been around for over a hundred years, where regular people connect, share, help, and explore the world through radio.

So when you hear someone call out on a repeater and you've got your license, don't hesitate. Pick up your radio, press the button, and say:

"This is [your callsign], just got licensed and listening in."

That's all it takes. You're part of the community now. And from here, the sky's the limit.

Chapter Quiz

How Much Do You Remember?

1. What's the name of the entry-level amateur radio license in the U.S.?

A) General License

B) Technician License

C) Operator License

D) Basic Radio Certificate

2. How many questions are on the Technician license test?

A) 25

B) 30

C) 35

D) 50

3. What score do you need to pass the test?

A) 50%

B) 60%

C) 75%

D) 90%

4. What tool lets you study the exact questions that might appear on the real exam?

A) Baofeng Radio App
B) FCC Exam Center
C) HamStudy.org
D) CallSign Generator

5. Which of the following is a benefit of getting your license?

A) Free radios
B) Transmitting legally on ham bands
C) International travel perks
D) Skip antenna setup

6. What's your first step before registering for an exam?

A) Buy a headset
B) Charge your radio
C) Get an FRN from the FCC
D) Take a Morse code test

7. You can legally talk on repeaters without a license.
True / False

8. The license is valid for 5 years once you pass.
True / False

9. You can take the Technician test online or in person.
True / False

10. After passing, you'll be assigned a callsign by the FCC.
True / False

_____

Answers: 1. B, 2. C, 3. C, 4. C, 5. B, 6. C, 7. False, 8. False, 9. True, 10. True

Talking Like a Licensed Pro

You've got your callsign. You've passed the test. You've earned the right to press that button and say something to the world, and now, for the first time, you can legally do it.

It's a big moment, and yeah, it might feel a little nerve-wracking at first. That's normal. Most people feel that tiny jolt of nerves before their first transmission. But remember...you're ready.

You've practiced. You've listened. You know how to use your radio. Now you're just doing it with confidence and full permission.

Let's walk through what your first transmission might sound like, and how to keep it smooth and simple.

Start by picking a quiet but active frequency. If you programmed a local repeater or a calling frequency like 146.520 MHz, that's a perfect place to begin. Turn to that channel, make sure your radio is set to transmit (not just listen), and wait a few seconds to be sure the frequency is clear.

When you're ready, press the PTT button and say:

"This is [your callsign], just got licensed and listening on 146.520."

Or, if you want to check if anyone can hear you:

"This is [your callsign], radio check please."

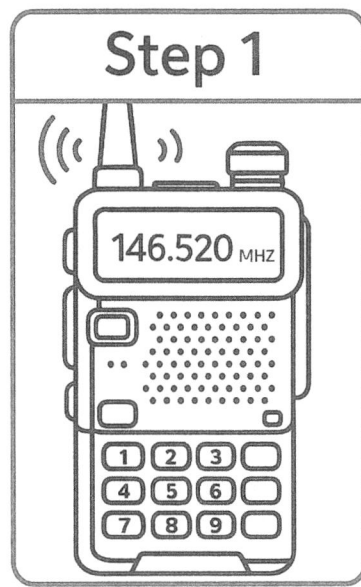
Tune your Baofeng to a calling frequency like 146.520 MHz

Press PTT and say your call sign and that you're listening

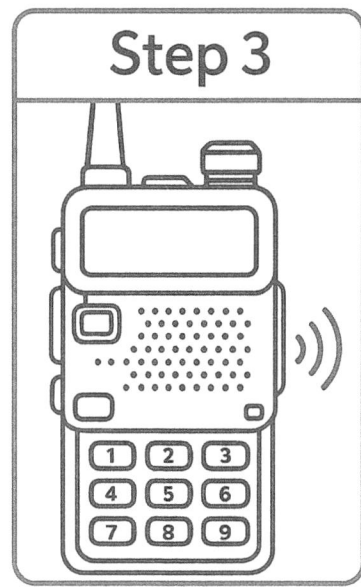
Release PTT, wait, and listen for someone to reply

Keep your voice calm, steady, and clear. Don't worry about sounding like a pro, just be yourself. When you're done speaking, release the PTT and listen.
If someone's around, they might come back with something like:

"KJ7ABC, I copy you loud and clear. Welcome aboard!"

And just like that, you're having your first real on-air conversation.

If no one answers, don't let it shake your confidence. It might just be a quiet moment. Try again later. Try a different frequency. Try in the evening when more people are on.

The important thing is, you showed up. You made your call. And next time, someone will be there to answer.

The first time you hear someone reply to your voice, it sticks with you. You'll remember the sound, the signal, and the moment your Baofeng wasn't just a tool anymore, it became a connection.

How to Find People to Talk To

You've keyed up the mic, made your first call, and maybe even gotten a response. That's a big win. Now you might be wondering, where is everyone? When are people actually on the air? And how do you find a good spot to tune in?

The truth is, radio activity can be a bit like fishing. Sometimes the bands are full of chatter. Other times, it's quiet. But there are a few simple ways to boost your chances of finding a conversation, and once you know where to look, it gets a lot more fun.

Let's start with the easiest and most reliable option: repeaters.

Repeaters are radio relays, usually set up high on towers or hills, that receive your signal and re-broadcast it over a much wider area. Think of them as megaphones for your handheld. They're a great place to find other operators, check into nets (organized radio meetups), or just hang out and listen to local traffic.

You can find repeaters near you using websites like repeaterbook.com. Just enter your zip code or city, and you'll get a full list of nearby repeaters, including:

- Their frequency (like 146.880 MHz)

- Whether they're open to the public (most are)

- What tone they use (if one's required for access)

- When activity is busiest (some list net times or club schedules)

Once you find a repeater that looks good, try programming it into your radio and listening for a few minutes during peak times, usually evenings and weekends.

Another great place to find people is on simplex calling frequencies. These are direct, one-to-one channels where people often meet before moving to a different frequency. On VHF, the national calling frequency is 146.520 MHz. On UHF, it's 446.000 MHz. These don't rely on repeaters, so they're perfect for line-of-sight chatting, especially in local areas or during events.

Want to find even more consistent activity? Try checking into a local net. These are regularly scheduled check-ins hosted by local ham clubs. They usually happen on the same repeater at the same time every week. Some are formal, others are casual, but either way, they're great for meeting friendly operators and learning the ropes. You can find local nets by:

- Visiting a local club's website

- Asking someone on-air

- Browsing repeater listings (many nets are listed with times)

- Checking sites like arrl.org or hamclubonline.com

And don't forget, radio is time-sensitive. Some frequencies are ghost towns at noon but hopping at 7 p.m. Try tuning in during:

- Early mornings (commuters checking in)

- Evenings (after dinner is prime radio time)

- Weekends (especially during contests or events)

- Stormy weather or power outages (you'll often find people sharing updates)

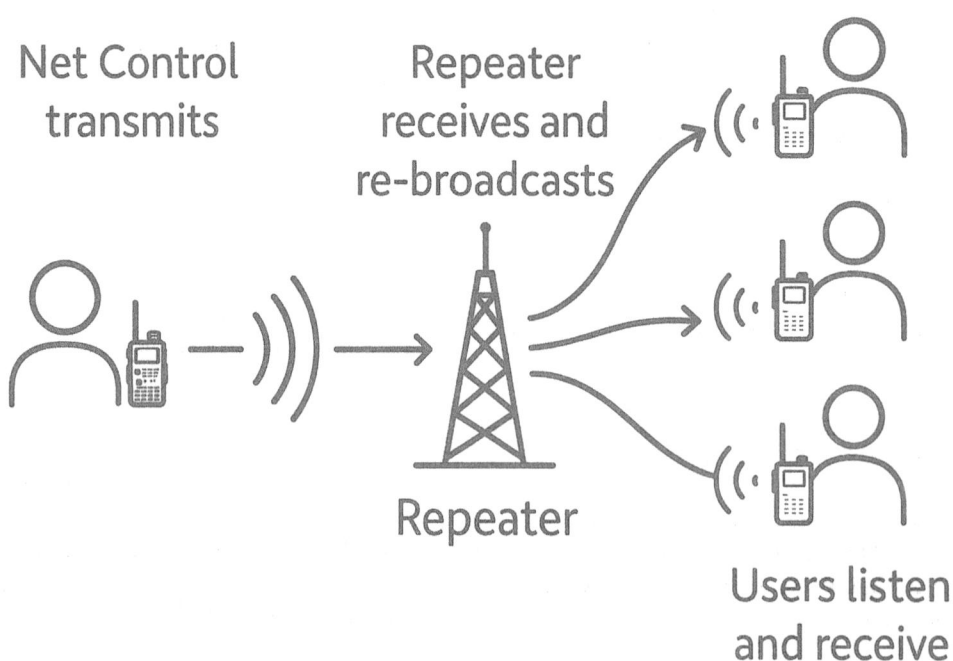

The more you explore, the more you'll get a feel for where the action is. Start by listening. Make a few calls. Try a net. Leave your radio on in the background while you're doing something else, you never know when you'll catch a voice calling out.

Making a Call: What to Say (And How to Say It)

You've found an active channel. Maybe you're tuned into a repeater or a quiet calling frequency. You're ready to talk. But now comes the big question: what do you actually say?

Good news, there's no secret handshake. You don't need fancy lingo or ham radio slang. All you need is a calm voice, your callsign, and a simple message.

Let's start with making an open call. This is where you put your voice out there to see if anyone's listening. Think of it like walking into a room and saying, "Hey, anyone want to chat?" Here's how to do it:

Step 1: Listen first. Give the channel 5–10 seconds to make sure it's not already in use.

Step 2: Press your PTT button, wait half a second, then say:

"This is [your callsign], monitoring [frequency or repeater name]. Anyone around?"

Step 3: Release the button and listen. Wait 10–20 seconds. If nobody answers, try again, maybe once or twice more. Then move on to another frequency or try again later.

That's a textbook "CQ" call, short, friendly, and to the point. You've identified yourself, said where you are, and invited a conversation.

Now, let's say someone answers and you want to respond. Here's how that works:

- Wait for their call to finish. Give them a second or two, then press your button.
- Say something like: "[Their callsign], this is [your callsign]. Good evening. You're coming in loud and clear."
- Then just let the conversation flow. Talk like you would in person, just a little slower and a little clearer.

If they ask where you are, what gear you're using, or how long you've been licensed, that's totally normal. You can say it's your first contact, most folks will be happy to hear it. This is a hobby full of people who remember exactly what that first moment felt like.

And what if you stumble a little? That's fine. You're not on stage. If you say something out of order or pause too long, just smile and say, "Still learning!" Most hams love helping newcomers feel more comfortable.

Once you've made a few one-on-one contacts, you might start hearing something a little different on the air, more organized, more regular, with more voices taking turns. That's called a net, and it's one of the most welcoming parts of the ham radio community.

A net (short for "network") is a scheduled on-air gathering. Think of it like a group meeting over the radio, where people check in, share news, and practice communication. Some nets are formal, others are casual. Some are for emergencies, others are just for fun. But they all follow a basic structure that makes it easy for anyone, even beginners, to jump in.

Here's how most nets work:

- There's a net control operator, someone who runs the show and calls the order.
- Operators are invited to "check in" one at a time, usually by giving their callsign.
- After everyone checks in, there may be short announcements, questions, or general conversation.
- Then the net wraps up, usually with a friendly thank-you and a group sign-off.

If you want to join a net, it's as simple as tuning in and listening. You'll often hear something like:

"This is [callsign], net control for the [club name] weekly net. All licensed stations are welcome to check in."

When it's your turn, press your PTT and say: *"This is [your callsign], no traffic, just checking in."*

That's it, you're in!

Some nets will ask if you have anything to share. Feel free to just say "no traffic" if you're there to listen and learn. Others might invite open conversation after the check-ins, and you can jump in if you'd like.

You can easily find nets by

- Checking local ham club websites
- Browsing RepeaterBook or QRZ.com listings
- Asking other operators on the air
- Listening around during early evenings, most nets happen then

If you're not quite ready to speak, that's okay, you can still listen to nets to get a feel for how they work. They're a great way to meet local operators, get comfortable with etiquette, and hear how real conversations happen.

Most nets are hosted on repeaters, those tower-mounted relay stations we talked about earlier. Repeaters help cover large areas, making them perfect for community check-ins. And now that you're licensed, you can use them anytime.

Joining a net, even just once a week, helps you stay active, connected, and confident. It's a simple reminder that you're part of something bigger, a network of people who know how to stay in touch, even when everything else goes quiet.

Logging Contacts and Starting Your Radio Log

Once you start talking to other operators, whether it's one-on-one or during a net, it's a good idea to keep track of who you've talked to, when, and on which frequency. This is called logging your contacts. Think of it like a journal for your radio activity. At first, it's a fun way to remember your first few conversations. Later on, if you jump into contests, make long-distance contacts (called DXing), or need to confirm who you've reached during an emergency, having a log becomes incredibly helpful. Here's what a basic radio log might include.

- Date and time of the contact
- The frequency or channel (like 146.520 MHz or a repeater ID)
- The callsign of the other operator
- Their location (if they mention it)
- A signal report (optional—how clearly you heard them)
- Any quick notes (like "first contact!" or "met through local net")

You don't need to log every conversation, but keeping track of your early contacts can help you see your progress

HAM RADIO CONTACT LOG

Date	Time	Frequency	Callsign	Notes
2025-04	19:15	146.520 MHz	Local net	First contact!
2025-04	19:30	146.880 MHz	Local net	Checked in, no traffic

Talking Like a Licensed Pro

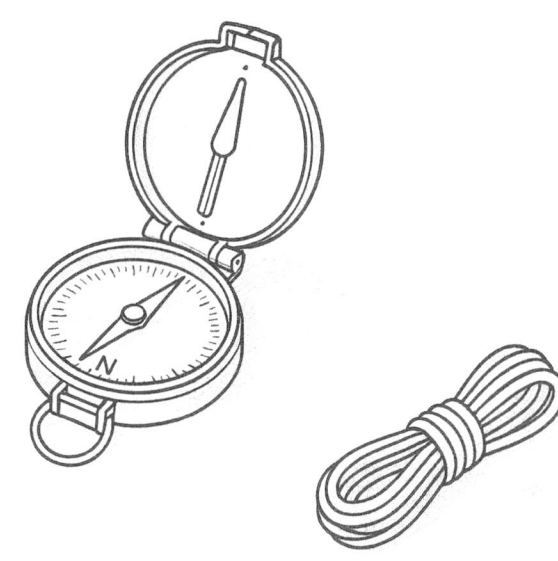

Handling Silence, Static, and Missed Calls

Sometimes you'll call out on a clear frequency, and... nothing happens.

Don't worry. That doesn't mean you did anything wrong. It just means no one was listening at that exact moment. Ham radio isn't like texting, there's no guaranteed reply. Sometimes it's just quiet. Sometimes your signal didn't reach far enough. And sometimes people are listening but simply not talking.

Here's what you can do when that happens:

- Try again in a few minutes. Wait a bit, then make another call.

- Switch frequencies. Move to a local repeater or another calling channel.

- Change times. Mornings, evenings, and weekends are often busier.

- Move to a better spot. Go outside or get higher up—line of sight helps a lot.

Silence is part of the game. The key is not to let it stop you. Every time you call out, you're getting more comfortable. And eventually, someone will answer.

You've learned how to program your radio, make calls, join conversations, and get comfortable on the air. That's already a huge achievement, but it's also just the beginning.

With your license and a little confidence, you can start exploring the full range of what ham radio has to offer. There's a whole world out there beyond local chatter, and your Baofeng is your gateway to it.

You might join a local emergency communication group, helping your community during wildfires, hurricanes, or power outages. Or you could take part in public service events, coordinating runners at a marathon or cyclists on a long-distance ride.

You might get into long-distance (DX) contacts, reaching operators across the country, or even around the world, with the right equipment. Some hams use radios to bounce signals off the moon, talk through satellites, or connect using digital modes that send messages like text across huge distances.

If you ever want to upgrade your license, you can study for the General or Extra Class, which unlock even more frequencies and long-range options.

But for now, the best thing you can do is simple, keep using your radio. Get on the air a little each week. Call out. Check into a net. Ask a question. Make a friend. Every contact builds skill, and every conversation makes this hobby even more rewarding. You're no longer just learning. You're a part of it.

Chapter Quiz

How Much Do You Remember?

_____

1. What should you always do before transmitting on a frequency?

A) Change your callsign

B) Increase power

C) Listen for a few seconds

D) Press scan

2. What's a good way to start your very first transmission?

A) Say 'Testing, testing!'

B) Say your name only

C) 'This is [your callsign], just got licensed and listening on 146.520.'

D) Ask for help immediately

3. What tool helps you reach more people by rebroadcasting your signal over a wider area?

A) Microphone booster

B) Repeater

C) Frequency scanner

D) Duplexer

4. Which of the following is considered good radio etiquette?

A) Talking nonstop for long periods

B) Interrupting other users

C) Keeping messages short and clear

D) Broadcasting music

5. What's the name of a scheduled on-air gathering where people check in and chat?

A) Talk thread

B) Channel roll

C) Signal loop

D) Net

6. If no one responds to your transmission, what should you do?

A) Assume your radio is broken

B) Keep calling every second

C) Try again later or switch frequencies

D) Yell louder

7. It's okay to call out without identifying yourself.
True / False

8. Simplex calling frequencies do not require repeaters.
True / False

9. You must check into a net using your full address.
True / False

10. Keeping a radio log can help track your progress and contacts.
True / False

_____

Answers: 1. C, 2. C, 3. B, 4. C, 5. D, 6. C, 7. False, 8. True, 9. False, 10. True

Master Your Radio's Hidden Tricks

By now, you've done the big stuff, you've powered on your radio, programmed your first channels, passed your license test, made your first calls, and started real conversations. That's your solid foundation.

And honestly? That's enough to use your Baofeng with confidence any day of the week.

But here's the fun part: your little handheld radio still has plenty more tricks up its sleeve.

You don't need to buy a new radio to start exploring more advanced features. You've already got the tools. It's just about unlocking them, one small step at a time. Whether you want to boost your range, save battery power, or stay better connected with a group, your Baofeng can do a lot more than you might think.

This chapter is all about leveling up, without the headaches. No complicated tech talk, no confusing menus. Just a few easy tweaks that help you get more out of the radio you already trust.

Here's some of the good stuff you're about to learn:

- How to listen to two channels at once (great for multitasking)
- How to boost your signal with a better antenna
- How to cut through repeater clutter using simple tones
- How to pick the right power setting for every situation
- How to stretch your battery life and make smart upgrades
- How to use your Baofeng more effectively for hiking, prepping, team events, and more
- Even how to copy your channels to another radio with just a few clicks

You don't have to master everything at once. Just pick one feature that sounds fun, try it out, and watch your skills grow.

Think of this chapter like tuning up a bike you already love riding, little upgrades that make the ride even smoother, faster, and way more fun.

Using Dual Watch and Dual Standby

Your Baofeng isn't limited to just one channel at a time. It can actually keep an ear on two different frequencies at once, flipping back and forth automatically. That means you can monitor your local net on one channel while also staying tuned to a calling frequency, an emergency alert, or even a buddy's walkie-talkie band, all without touching a button.

This feature is called dual watch or dual standby, and it's built right into your Baofeng. You've probably noticed your screen shows two lines, one on top, one on the bottom. Each one can be set to a different frequency or channel. When dual watch is turned on, your radio listens to both and stops on whichever one has activity first.
Here's how to turn it on:

Step 1: Make sure you're in Channel Mode. Press the VFO/MR button if you're still in Frequency Mode.

Step 2: Set one channel on the top line (A) and another on the bottom line (B). Use the arrow keys to scroll through your saved channels.

Step 3: Press the MENU button, scroll to TDR (Menu #7 on most models), and press MENU again. (TDR stands for "Transmit Dual Reception.")

Step 4: Set TDR to ON, then press MENU again to confirm.

Step 5: Press EXIT. That's it, you're now in dual watch mode.

Now your radio will flip between both channels and stop when it hears activity on either one. The A/B button lets you choose which channel you'll transmit on when you press PTT, so make sure the one you want to reply on is selected.

It's a simple way to stay alert and responsive across two zones of communication, whether you're using saved channels or typing in frequencies manually.

Quick Tip: If you're hearing too much chatter, or if your radio keeps flipping between channels too often, you can turn dual watch off the same way you turned it on, just set TDR to OFF.

Switching Transmit Power for Better Range or Battery Life

Your Baofeng has two power settings: high and low. High power gives you more reach, especially when you're trying to hit a distant repeater or operating outdoors where signals don't bounce easily. Low power saves battery life, making it perfect for short-range chats, indoor use, or monitoring channels all day without draining your radio.

Quick Tip: High power drains your battery faster, usually giving you about 5–6 hours of heavy use. Low power can stretch your battery life to 10 hours or more, especially if you're mostly listening.

Switching between high and low power is quick and easy, and once you know when to use each one, you'll be able to make smarter choices depending on your situation. Here's how to adjust your transmit power.

Step 1: Press the **MENU** button.

Step 2: Scroll to **TXP** (usually Menu #2). This stands for "Transmit Power."

Step 3: Press **MENU** again to select it.

Step 4: Use the arrow keys to choose either **HIGH** or **LOW**.

Step 5: Press **MENU** to confirm, then **EXIT** to return to the main screen.

Now, your radio will transmit at the power level you chose, depending on the channel you're using. So, when should you use each power setting?

Use HIGH power when:
- You're trying to reach a repeater that's far away.
- You're outdoors in a wide-open area with fewer signal reflections.
- You're testing your signal strength or doing a radio check with someone at a distance.

Use LOW power when:

- You're talking to someone nearby, like inside the same building or at the same park.
- You're checking into a local net on a strong, nearby repeater.
- You want to conserve battery life, especially during long-term listening.
- You're practicing or testing indoors, where high power isn't really needed.

Improving Your Signal With Antenna Upgrades

One of the fastest, easiest, and most effective upgrades you can make to your Baofeng is swapping out the stock antenna. The one that came in the box? It'll do the job, but it's basic. Think of it like the tiny spare tire that comes with your car: it'll get you moving, but it's not built for real performance.

A better antenna helps your radio send and receive clearer signals over longer distances. It won't magically double your range, but you'll absolutely notice the difference, especially when you're trying to reach a repeater or talk to someone in a noisy area.

The best part? Swapping it out takes about 30 seconds. Just unscrew the stock antenna and screw the new one in, it's that easy. Here are a few antennas I recommend for beginners:

Nagoya NA-771: A long, flexible whip antenna that gives a noticeable boost on both VHF and UHF bands. Perfect for general use and outdoor conditions.

Nagoya NA-701: Shorter and stiffer than the 771, but still a huge upgrade from the original. Great for indoor use or if you want something lighter on your belt or backpack strap.

Signal Stick (by Signal Stuff): A flexible, super-durable antenna designed by ham radio volunteers. It performs great, weighs next to nothing, and bends instead of breaking.

When you're shopping for an antenna, make sure it uses an SMA-Female connector, which fits most Baofeng models like the UV-5R and BF-F8HP. If you're using a different model, double-check the connector type before ordering, just to be safe.

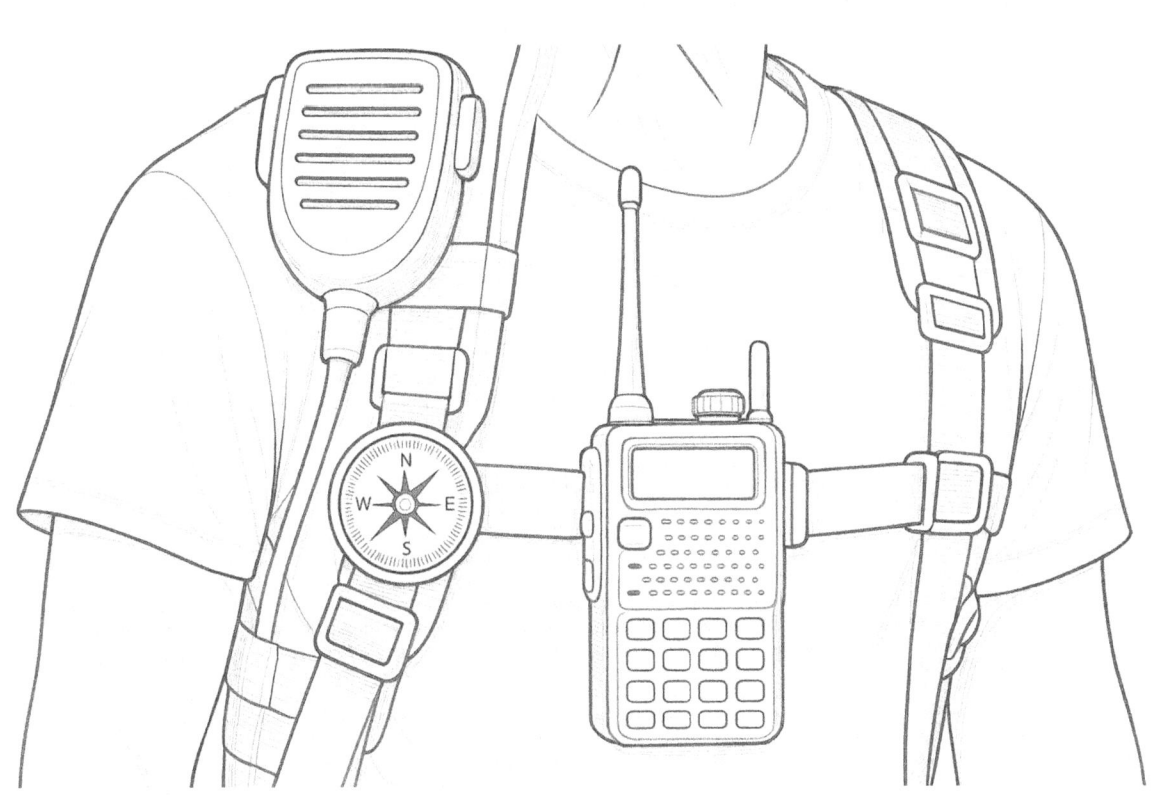

After you swap antennas, you don't need to tweak anything else. Your radio will work exactly the same, just with better reach and reception.

If you've been struggling to hear someone clearly, or your signal keeps dropping out, upgrading your antenna is the first thing to try. It's one of the easiest, fastest wins you can get with a Baofeng.

Quick Tip: If you only grab one upgrade, make it the antenna. A simple $20 whip can make your Baofeng feel like a whole new radio.

Understanding CTCSS and DCS for Repeaters

If you've ever tried to reach a repeater and heard nothing, or got a weird buzzing sound instead of a clear connection, you've probably run into something called a CTCSS or DCS tone. Don't let the acronyms scare you, here's the plain-English version.

Some repeaters need a little "password tone" before they'll let your signal through. It's not a secret code, it's just a low-frequency sound your radio sends along with your voice. If the repeater hears the right tone, it opens up. If it doesn't, nothing happens.

These tones help keep the airwaves clean by blocking accidental signals, interference, or radios that aren't set up correctly.

There are two types:

CTCSS (Continuous Tone-Coded Squelch System): Think of it like a soft, background "ping" that opens the door to the repeater.

DCS (Digital-Coded Squelch): Same basic idea, but digital. It's less common for beginners, but it works the same way.

Most repeater listings *(on sites like repeaterbook.com)* will tell you exactly what tone you need. You'll usually see something like:

146.880 MHz, PL 103.5

That "PL" number is the CTCSS tone, just plug it into your radio, and you're good to go. Here's how to set it up on your Baofeng:

Step 1: Go to the frequency you want to use (make sure you're in Frequency Mode).

Step 2: Press MENU, then scroll to T-CTCS (usually Menu #13).

Step 3: Press MENU again, then use the arrow keys to select the tone (like 103.5).

Step 4: Press MENU to confirm, then EXIT.

Optional Tip: You can also set R-CTCS if you want your radio to only open up when it hears a signal with the same tone. This is super handy in noisy areas where random static might otherwise break through.

Once you've got the tone entered, try transmitting again. If it's the right one, the repeater should respond, and your voice will reach a much wider area.

The good news? You only need to set the tone once per repeater. You can save it as part of a channel using CHIRP, or the manual save method we covered earlier. It's a small tweak that unlocks a big part of your radio's true potential.

Better Accessories

You've got the radio, you've got the license, and you've even figured out the features. Now it's time to talk about the extras, simple accessories that can make your Baofeng easier, louder, longer-lasting, and just plain nicer to use.

These aren't must-haves, but they can make a big difference, especially if you're using your radio regularly, on the move, or in noisy environments. Here are a few of the best ones to consider:

Speaker Microphone. Clip this to your shirt or backpack strap and talk without needing to pull your radio off your belt or bag. It's perfect for hiking, group events, or working outdoors. Just plug it into the side port, and you're good to go.

High-Capacity Battery Pack. Most Baofengs come with a standard battery, but you can upgrade to a larger one that lasts twice as long. It's great for emergency kits, long hikes, or all-day events when recharging might not be an option.

USB Charging Cable or Car Charger. Some models and aftermarket kits include USB charging adapters. These let you charge your radio from a car, power bank, or laptop, making them super handy for travel and on-the-go setups.

Belt Clip or Holster Case. A simple clip keeps your radio on your hip or backpack strap for quick access. If you're active or outdoors a lot, a padded holster helps protect the screen and body while keeping everything secure.

Earpiece with Push-to-Talk Button. An earpiece is perfect when you want to stay quiet or go hands-free, whether you're working an event, biking, or listening during a group outing. Some models even have a PTT button built into the wire, so you don't have to reach for your radio.

Programming Cable (If You Didn't Get One Yet). It's worth repeating, get a good-quality programming cable if you plan to use CHIRP. It'll save you hours of button-pushing and make managing your channels fast and easy.

Most of these accessories cost under $25 and are easy to find online. You don't need to grab everything at once, just start with what fits your needs right now and build from there.

Cloning Channels With CHIRP (The Easy Way)

Imagine you've programmed your Baofeng with all your favorite channels, repeaters, weather stations, local nets, emergency frequencies, and now you want another radio set up exactly the same way. Maybe it's for a backup, a family member, or a teammate on a group hike.

Good news, you don't have to reprogram everything by hand. With CHIRP, you can copy your entire channel list from one radio to another in just a few minutes. It's fast, easy, and makes sure both radios are on the same page, ready to go.

Step 1: Connect your original radio to your computer with the programming cable. Open CHIRP, click on Radio > Download From Radio, and load up your current settings.

Step 2: Save the file to your computer. Go to File > Save As and give it an easy-to-remember name like "Hiking Setup" or "Main Radio Config."

Step 3: Disconnect the first radio and plug in the second one. Power it on, open CHIRP again, and this time click Radio > Upload To Radio.

Step 4: Select the file you saved, double-check the radio model, and hit Upload. CHIRP will load all your saved channels, tones, names, and power settings into the second radio.

And that's it, you've just cloned your setup! This is especially handy for:

- Setting up radios for group outings
- Prepping a backup radio for emergencies
- Giving a friend the same channel list so you can talk right away
- Testing new setups without overwriting your main radio

You can even create multiple profiles for different situations, one for hiking, one for around town, one for family events, and swap them in and out with CHIRP in just seconds.

Now that your radios are synced up, let's wrap up this chapter with some real-world examples of how to use your Baofeng more effectively, depending on where you are and what you're doing.

Using Your Radio in Real-Life Scenarios

Now that you've mastered the basics (and picked up a few advanced tricks), let's talk about putting your Baofeng to work in real-life situations. This is where all the settings, accessories, and good habits you've built start to really pay off.

Here are some common scenarios where your radio can shine, and a few tips for getting the most out of it in each one.

Hiking or Backcountry Trips. Miles into a trail with no cell service? Your Baofeng keeps you connected with your hiking partner, or lets you call for help if needed. Use dual watch to monitor a calling frequency and your team's channel at the same time.

Keep your radio on low power to conserve battery, and always pack a spare or high-capacity battery just in case.

Power Outages or Natural Disasters. When the power's out and cell towers are down, your radio still works. Tune into your local NOAA weather station, monitor emergency channels, or check in on a local repeater net.

Keep your radio charged and ready with a car adapter or USB cable. This is exactly when all that quiet practice pays off.

Group Events or Family Outings. Heading to a fair, a hike, or a big group trip? Clone your channel list to other radios and stay connected without relying on phones. Use earpieces or speaker mics to keep communication hands-free and easy.

Road Trips or Convoys. Driving with a group? Skip the static of CB radios and use your Baofengs for clean, direct communication. Program FRS/GMRS-compatible channels or a shared simplex frequency. Clip your radio to the dash or seatbelt strap with a speaker mic for easy, hands-free talking.

Training, Volunteering, or Public Service. Helping out at a race, marathon, or public event? Bring your radio and offer to assist. Many volunteer groups use handhelds like yours to relay updates, check on participants, and coordinate logistics.

These are just a few examples, but they highlight the bigger point, your Baofeng isn't just a toy. It's a real tool with real purpose. And now that you know how to use it well, you're not just prepared, you're ahead of the curve.

Chapter Quiz

How Much Do You Remember?

___

1. What feature lets your Baofeng listen to two channels at once?

A) Dual Scan

B) TDR (Dual Watch)

C) Split Mode

D) Cross Band Receive

2. When should you use LOW transmit power?

A) Talking to someone nearby

B) Long-distance emergency transmission

C) During contests

D) When you want maximum range

3. Which antenna is recommended for boosting your signal over the stock one?

A) Stock stubby

B) Loop whip

C) Nagoya NA-771

D) Shortwave stub

4. What does CTCSS do for repeater communication?

A) Boosts your volume
B) Makes your signal invisible
C) Adds a tone to access certain repeaters
D) Encrypts your message

5. Which accessory allows hands-free use while walking or biking?

A) Car charger
B) Speaker mic
C) USB adapter
D) Earpiece with PTT

6. What's one real-life reason to clone your channels to another Baofeng?

A) To reset your settings
B) For software updates
C) To keep two radios on the same setup
D) To avoid squelch errors

7. High transmit power always gives better range and should be used at all times.
True / False

8. Dual Watch allows your radio to flip between two active channels.
True / False

9. All antennas use the same connector, so any will fit your Baofeng.
True / False

10. Keeping a charged spare battery is smart for hiking or emergencies.
True / False

Answers: 1. B, 2. A, 3. C, 4. C, 5. D, 6. C, 7. False, 8. True, 9. False, 10. True

Advanced Features for Everyday Use

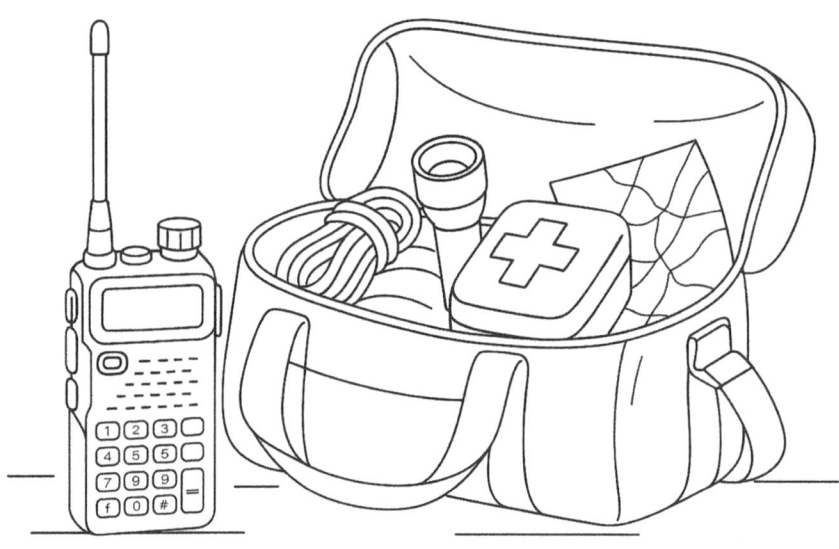

By now, you know how repeaters work. They take your signal, boost it, and send it back out so people farther away can hear you. It's like putting your voice on a loudspeaker mounted on a mountain.

But here's a small technical detail that trips up a lot of beginners: you don't transmit and receive on the exact same frequency when you're using a repeater.

Instead, repeaters use something called an offset, meaning the input (what you transmit on) and the output (what you listen to) are slightly different. Your Baofeng needs to know about this offset so it can talk to the repeater properly.

Offset

Let's say you want to talk on a repeater with an output frequency of 146.880 MHz. That's what you tune in to hear others. But to talk back, your radio needs to transmit on 146.280 MHz instead. That's a 0.6 MHz offset in the minus direction.

Here's what you need to know:

- On the 2-meter band (VHF), most repeaters use a 0.600 MHz offset.

- On the 70-centimeter band (UHF), the offset is usually 5.000 MHz.

- The direction can be + (plus) or − (minus), depending on the repeater.

Don't worry, most Baofeng radios can handle this automatically once you set the shift direction and offset value.

How to Set Offset and Shift on Your Baofeng

Here's how to do it manually if you're programming a repeater directly into your radio:

Step 1: Enter Frequency Mode (VFO) Press the VFO/MR button until you see a frequency, not a channel name.

Step 2: Enter the repeater's output frequency (the one you listen to) Example: 146.880 MHz.

Step 3: Set the offset value Press MENU, scroll to OFFSET (usually Menu #26), press MENU again, then enter 0.600 for VHF. Press MENU to confirm.

Step 4: Set the shift direction Go to SHIFT-D (Menu #25). Choose +, –, or OFF depending on what the repeater uses. Press MENU to confirm.

Now, when you press the PTT button, your Baofeng will automatically transmit on the correct input frequency, without you needing to switch anything manually.

Where Do You Find the Offset Info?

Repeater listings (like on repeaterbook.com) will usually show something like: **146.880 MHz, – Offset, PL 103.5**

That tells you the output frequency (what you listen to), the shift direction (minus), and the tone (PL/CTCSS) if needed.

Quick Tip: If you're using CHIRP, you can enter all of this information into one channel slot, frequency, tone, offset, and power level, and save it with a clear name like "Local Repeater" so it's easy to find later.

Priority Channel Monitoring

Imagine you're scanning through a list of channels, maybe local repeaters, weather frequencies, and simplex channels, but you always want your radio to keep one ear tuned to a specific one. Maybe it's the weather station. Maybe it's your team's coordination channel. Or maybe it's a local repeater that gets busy during storms. That's where a priority channel comes in.

On many radios, you can set one channel to act like a "home base." Even while scanning, your radio keeps checking that priority channel every few seconds. If it hears something there, it automatically jumps to it and stops scanning.

Now, your Baofeng doesn't have a true "priority channel" button like some high-end models, but you can easily simulate this using dual watch (TDR) with a quick setup.

Step 1: Turn on Dual Watch (TDR) Press MENU, scroll to TDR (Menu #7), and set it to ON.

Step 2: Set your priority channel on one line (A) For example, set line A to 146.520 (the national simplex calling frequency) or a local repeater you want to monitor constantly.

Step 3: Set the other line (B) to the channel you want to scroll through or scan

Step 4: Select which line will transmit Use the A/B button to choose which channel your PTT button will use when you talk.

Now, while you're scanning or chatting on the B channel, your Baofeng will still listen for activity on your A channel. If someone starts transmitting there, the radio will automatically switch over so you don't miss it. This is great for

- Monitoring a local emergency channel while scanning other bands.
- Keeping an ear on your group's main frequency during a hike or event.
- Listening for weather alerts while still participating in a net.
- Staying ready during power outages or storm conditions.

Dual watch makes your radio feel like two tools in one, and once you get the hang of it, you'll wonder how you ever lived without it.

Wideband vs. Narrowband: What It Is and When to Use It

As you explore more Baofeng features, you'll come across a setting labeled W/N, short for Wideband/Narrowband. It might sound complicated, but it's really just a simple tweak that changes how your voice sounds and how much space your signal uses on the airwaves.

Advanced Features for Everyday Use

Wideband (WIDE): Sends a stronger, fuller-sounding signal that uses more of the frequency. It's also great for clear voice chats, but it takes up more "room" and can crowd other users. Stick to Wideband when:

- You're chatting simplex (radio-to-radio) without a repeater.
- You're in a wide-open area with little radio traffic.
- You want the strongest, clearest voice quality and nobody minds a wider signal footprint.

Narrowband (NARROW): Sends a slightly smaller, tighter signal that uses less bandwidth. It's a little quieter, but perfect for crowded areas or repeaters that require it.

- A repeater or club specifically asks for it.
- You're operating close to others and want to avoid causing interference.
- People tell you your audio sounds too loud or distorted.
- You're coordinating with narrowband-only radios, like FRS or commercial radios.

Here's how to change this setting on your Baofeng

Step 1: Press the MENU button, then scroll to W/N (usually Menu #5).

Step 2: Press MENU again to select it.

Step 3: Use the arrow keys to choose either WIDE or NARR.

Step 4: Press MENU to confirm, then press EXIT to return.

Each saved channel can have its own setting, so you can mix and match depending on where and how you use your radio.

*Quick Tip: If you're not sure which one to use, start with **WIDE** for general chatting or simplex conversations. If someone tells you your audio sounds a little harsh or distorted, try switching to **NARROW** and see if it improves.*

VOX: How to Use Voice-Activated Transmission

Ever wish you could talk into your radio without pressing the PTT (Push-To-Talk) button? That's exactly what VOX mode is for. VOX stands for Voice-Operated Exchange, and when it's turned on, your Baofeng will automatically start transmitting whenever it hears your voice.

That means totally hands-free operation, perfect for hiking, biking, working on a project, or anytime you want to stay in touch without juggling buttons.

When Is VOX Useful?

- During bike rides or hikes, when pressing a button isn't practical
- While working on projects like rooftop repairs, garage work, or trail maintenance
- As part of a go-bag or emergency kit for fast, hands-free communication
- For disability-friendly operation, when pressing buttons might be difficult

What to Know Before You Use It

- Your radio will transmit any sound it picks up, not just your voice. Background noise like wind, car engines, or nearby conversations can accidentally trigger it.
- Always test VOX in a quiet place first to adjust the sensitivity properly.
- If you're in a noisy area, it's better to stick with the PTT button or use an earpiece with a mic for better control.

How to Turn VOX On and Adjust Sensitivity

Step 1: Press MENU, then scroll to VOX (usually Menu #4).

Step 2: Press MENU again to select it.

Step 3: Choose a sensitivity level from 1 (least sensitive) to 10 (most sensitive).

Try starting around level 3 to 5, and adjust based on your environment.

Step 4: Press MENU to confirm, then press EXIT.

To turn VOX off, just return to the VOX menu and set it to OFF.

Once VOX is activated, speak into the mic. After a short pause, you'll hear the familiar "beep," and your radio will start transmitting automatically. Stop talking, and after a second or two, it will go back to listening.

It might take a little practice to get used to VOX, but once you do, it's a powerful tool, especially when hands-free operation means extra safety or convenience.

Smarter Scanning

Your Baofeng can automatically scan through all your saved channels, listening for any activity and stopping when it picks something up. That's what channel scanning is built for. But without a few simple tweaks, scanning can feel… well, a little chaotic.

It might stop on dead air, pause on random static, or zip right past the channels you actually care about. The good news? With a few smart adjustments, you can make scanning much more useful, and way less frustrating.

Scanning can be a superpower, or a source of endless static, depending on how you set it up. Here's how to make it work better for you:

1. Skip Empty Channels. Your Baofeng lets you mark channels to skip during scanning. This is super helpful if:

- You've saved placeholder channels you don't use yet
- A frequency always has static or dead air
- You only want to scan a smaller set of important channels

How to skip a channel:

Step 1: Go to the channel you want to skip (in Channel Mode)

Step 2: Press MENU, then scroll to SCAN ADD (usually Menu #30)

Step 3: Press MENU again, and set it to DEL (to remove it from the scan list)

Step 4: Press MENU to confirm, then EXIT

(Need to re-add it later? Just go back and set it to ADD.)

2. Start a Scan Once you've cleaned up your channel list, it's time to start scanning:

- In Channel Mode, press and hold the SCAN button (often the bottom left key marked "" or "SCAN")
- Your radio will scroll through your saved channels and stop when it hears something
- If it picks up nothing after a few seconds, it'll keep moving
- Press any key to stop scanning manually

3. Keep Your List Clean

- Remove test or temporary channels you don't really need
- Group related channels together (like emergencies, repeaters, simplex) so you can scan smarter
- Use CHIRP to manage your scan list—it's way faster and easier than doing it all through the radio buttons

Organizing and Labeling Channels

Once you've saved a few channels to your radio, repeaters, weather updates, calling frequencies, or team channels, it doesn't take long before you start thinking, *Was CH-012 my hiking frequency, or was that the net check-in?*

That's where smart channel organization comes in. Labeling and grouping your channels isn't just about being neat, it makes scanning faster, switching channels easier, and using your radio in real-life situations way less confusing.

On the Baofeng itself, you can't rename channels directly, you'll always see CH-001, CH-002, and so on. But when you use CHIRP, the free programming software, you can assign each channel a short name that shows up when you scroll through your radio's screen. Instead of just numbers, you'll see something useful like "REPEATER1," "NOAA WX," or "HIKE COM."

To label your channels, open your saved radio file in CHIRP and look for the "Name" column next to each frequency. You can type in up to 7 or 8 characters, depending on your model. Keep the names short, clear, and easy to recognize at a glance.

Here are a few simple naming ideas that work well:

- **CALLING** — for 146.520 or 446.000
- **REPTR1 / REPTR2** — for your local repeaters
- **WX NOAA** — for weather broadcasts
- **TEAM1 / HIKING** — for event or group channels
- **NET MON** — for weekly net check-ins
- **FAM CH1** — for FRS-style family communication
- **SCAN SKP** — for channels you want to test but might want to skip during scanning

You can also group your channels by function. For example:

- **CH 001–010:** Repeaters
- **CH 011–020:** Simplex and calling frequencies
- **CH 021–030:** Emergency and weather channels
- **CH 031–040:** Custom or event-based channels

Organizing your channels like this makes scanning and switching way easier. When you're flipping through, you'll know exactly where to find what you need, no more guessing or endless scrolling.

As your channel list grows, it's easy to lose track of what you've already saved. That's why it's smart to save a backup copy of your CHIRP setup every time you make changes. Just give it a clear, easy-to-remember name, like *"Baofeng_HikingSetup_April2025"*, so you can reload it anytime you need to.

Organizing your channels isn't about being fancy, it's about being ready. Whether you're checking the weather during a storm or staying connected with friends on a trail, good labels and a clean channel list help you move faster, stay confident, and avoid mistakes when it matters most.

Hidden Extras That Might Actually Come in Handy

Baofeng radios aren't just built for serious communication, they also come with a few hidden tools that can really save the day. You might've already spotted them and thought, "Why does my radio have a flashlight?" or "Wait, it plays FM radio too?" These little features might seem silly at first, but in an emergency or out on the trail, they can be incredibly useful.

Let's start with the FM radio. This isn't for talking to other people, it's the same kind of FM you'd hear in your car or a portable stereo. Your Baofeng can tune into local broadcast stations, letting you hear music, news, or emergency updates. That might not seem like a big deal, until the power goes out and your phone has no service. Suddenly, being able to hear a news update becomes priceless.

To use the FM radio, press the **MONI** button (sometimes labeled "Call" or "FM" on certain models). Use the arrow keys to scroll through stations, and press the button again to exit when you're done.

And then there's the flashlight. Is it a super bright beam? Nope. But if your headlamp dies or your phone flashlight gives out, that tiny white LED on top of your Baofeng is enough to find your keys, check a map, or dig something out of your bag at night. To turn it on, just press and hold the button just under the PTT. Tap it again to turn it off.

These little extras aren't why you buy a Baofeng, but they're a big part of why you'll be glad you have one. It's another reason your Baofeng belongs in your bag, your glove box, your go-kit, or even clipped to your backpack when you head out the door.

Chapter Quiz

How Much Do You Remember?

___

1. What does 'offset' refer to when using a repeater?

A) The size of your antenna

B) The color of your display

C) The difference between transmit and receive frequencies

D) The tone needed to access a channel

2. What's the usual VHF offset value used by most repeaters?

A) 0.300 MHz

B) 0.600 MHz

C) 5.000 MHz

D) 1.000 MHz

3. What does enabling TDR (Menu #7) on your Baofeng allow you to do?

A) Use the flashlight

B) Scan faster

C) Monitor two frequencies at once

D) Program channels with CHIRP

4. What's a real-world reason to use VOX mode?

A) Louder audio

B) Hands-free talking while hiking

C) Better squelch

D) Access encrypted channels

5. Which accessory helps you operate your radio without pulling it off your belt?

A) Longer antenna

B) USB cable

C) Speaker microphone

D) Programming cable

6. What is 'Wideband' typically best for?

A) Saving battery life

B) Narrow channels

C) Clear simplex voice chats

D) Muting background noise

7. The Baofeng flashlight can help in low-light situations.
True / False

8. You can rename channels directly on the Baofeng without CHIRP.
True / False

9. SCAN ADD menu lets you skip channels during scanning.
True / False

10. You can use FM radio on your Baofeng to listen to local music and news.
True / False

_____

Answers: 1. C, 2. B, 3. C, 4. B, 5. C, 6. C, 7. True, 8. False, 9. True, 10. True

Finding Your Radio Rhythm

You've made it. You unpacked your first Baofeng, programmed real channels, scanned the airwaves, passed your license exam, and made your first calls. You've even tweaked settings most people never touch, learned how to talk like a pro, and started exploring some powerful tools.

You're not guessing anymore, you're operating. Now comes the next big question, **how do you keep growing from here?**

The good news is, you don't have to do anything fancy. You don't need special gear or big projects. All you need is a rhythm, a simple, steady habit of using your radio a little bit at a time. That's how real confidence builds. That's how you go from "I'm learning" to "I've got this."

So, what does that rhythm look like?

Here's a good place to start:

Check in once a week.

Turn on your radio and scroll through a few channels. Call out, even if it's just once. If no one responds, that's okay, you're building the habit.

Listen to a net.

Even if you're not ready to talk yet, listening teaches you more than you realize. You'll start recognizing familiar voices, understanding the flow of conversations, and feeling more at home on the air.

Try one small thing each month.

Maybe it's testing a new antenna, joining a local event, checking out a digital mode, or just organizing your channel list with CHIRP. One small step is enough.

Keep your radio charged and within reach.

That simple choice turns your Baofeng from a gadget into a real tool, ready for emergencies, ready for fun, ready for anything.

Stay curious.

Follow a new ham YouTube channel. Join a Facebook group. Visit a local club. Ask questions. Help someone else once you feel ready. Staying curious is how you stay plugged in.

You don't need to chase every feature or buy new gear every month. You don't need to be the busiest operator on the air. You just need to keep showing up, keeping your gear ready, and using your radio in ways that work for you.

There's no finish line in this hobby, just the next transmission, the next conversation, the next moment you reach for your Baofeng and feel ready.

You're not a beginner anymore.

You're a radio operator.

You're radio-ready.

Book Quiz

How Much Do You Remember?

_____

1. Which button lets you switch between Frequency Mode and Channel Mode?

A) MENU

B) VFO/MR

C) A/B

D) PTT

2. What's a benefit of using a repeater?

A) Adds encryption

B) Increases range

C) Uses less battery

D) Improves weather alerts

3. Which of these should you do BEFORE pressing the PTT button?

A) Change the antenna

B) Lower the volume

C) Check the frequency is clear

D) Set TXP to LOW

4. What does a CTCSS tone help with?

A) Better scanning
B) Privacy settings
C) Accessing certain repeaters
D) Programming faster

5. What's a common VHF calling frequency?

A) 462.7125 MHz
B) 146.520 MHz
C) 162.500 MHz
D) 446.000 MHz

6. What menu setting changes how wide your signal sounds?

A) SQL
B) W/N
C) TDR
D) SCAN

7. What does 'offset' refer to in repeater programming?

A) Audio balance
B) Battery calibration
C) Transmit/receive frequency difference
D) Antenna length

8. Why would you use LOW transmit power?

A) To extend signal range
B) For long-range testing
C) To conserve battery
D) To scramble your voice

9. What does VOX mode allow you to do?

A) Control volume remotely
B) Transmit by voice without pressing PTT
C) Block interference
D) Silence squelch

10. Which of these is a good first step for organizing channels?

A) Add encryption
B) Rename them with CHIRP
C) Set all to HIGH power
D) Erase old settings

11. What's the function of the SCAN ADD menu?

A) Delete channels
B) Boost receive volume
C) Include/exclude channels from scanning
D) Program VOX

12. What can a speaker mic help with?

A) Better battery life
B) Avoiding squelch
C) Hands-free talking
D) Programming CHIRP

13. What kind of antenna connector do most Baofengs use?

A) SMA-Female
B) BNC
C) USB-C
D) SMA-Male

14. What's a typical UHF offset?

A) 0.6 MHz
B) 2.5 MHz
C) 5.0 MHz
D) 1.0 MHz

15. Where can you find local repeater info?

A) Your Baofeng manual
B) QRZ.com
C) Repeaterbook.com
D) CHIRP

16. What's one reason your call may not get answered?

A) Wrong CTCSS
B) You forgot to press EXIT
C) Frequency was too narrow
D) Antenna was too short

17. What does CHIRP allow you to do easily?

A) Increase transmit power
B) Scan faster
C) Program and clone channels
D) Change squelch automatically

18. What does the TDR menu do?

A) Boost volume
B) Enable dual watch
C) Lock keypad
D) Tune FM radio

19. What's a reason to use Narrowband (NARR)?

A) More audio clarity
B) Quieter receive
C) Less signal interference
D) Improved range

20. Which channel naming system helps with fast access?

A) Numbering by day
B) Using emojis
C) Grouping by function
D) Random order

21. Which frequency is best for NOAA weather alerts?

A) 446.000 MHz
B) 462.5625 MHz
C) 146.520 MHz
D) 162.550 MHz

22. What's the max number of memory channels on most Baofengs?

A) 64
B) 99
C) 128
D) 200

23. What's a sign that your squelch is too high?

A) Battery drains fast
B) You miss weak signals
C) Screen flickers
D) Audio is distorted

24. What's a good first message after getting licensed?

A) 'Breaker breaker!'
B) 'Testing!' (repeatedly)
C) 'This is [callsign], just got licensed and listening.'
D) Silence

25. When cloning with CHIRP, what step comes AFTER saving the first radio's file?

A) Erase memory
B) Upload to second radio
C) Rescan channels
D) Adjust squelch

26. Dual watch allows your radio to monitor two channels at the same time.
True / False

27. VOX mode only activates with your voice, not background noise.
True / False

28. A narrowband signal takes up more space than a wideband one.
True / False

29. You can use CHIRP to save backups of your programmed channel list.
True / False

30. All repeaters require a CTCSS or DCS tone to access.
True / False

Answers: 1. B, 2. B, 3. C, 4. C, 5. B, 6. B, 7. C, 8. C, 9. B, 10. B, 11. C, 12. C, 13. A, 14. C, 15. C, 16. A, 17. C, 18. B, 19. C, 20. C, 21. D, 22. C, 23. B, 24. C, 25. B, 26. True, 27. False, 28. False, 29. True, 30. False

Appendix A: Common Frequencies for Beginners

Here are some reliable, beginner-friendly frequencies you can program into your Baofeng right away. These are commonly used across the U.S. and are perfect for practicing, listening, or connecting with others.

Simplex Calling Frequencies (Direct Radio-to-Radio)

- 146.520 MHz – National 2m (VHF) calling frequency
- 446.000 MHz – National 70cm (UHF) calling frequency

NOAA Weather Broadcasts (Receive Only)

- 162.400 MHz
- 162.425 MHz
- 162.450 MHz
- 162.475 MHz
- 162.500 MHz
- 162.525 MHz
- 162.550 MHz

(Note: Frequencies vary by location. Try each one to find your local broadcast.)

FRS/GMRS Channels (For Family/Group Use) (Listen-only if unlicensed for GMRS)

- Channel 1 (462.5625 MHz)
- Channel 3 (462.6125 MHz)
- Channel 5 (462.6625 MHz)
- Channel 7 (462.7125 MHz)

Local Repeaters Look up your area on repeaterbook.com and program:

- Output frequency (what you listen to)
- Input offset and shift direction (what your radio transmits)
- PL or CTCSS tone, if required

Appendix B: Baofeng Menu Guide

Menu #	Name	What It Does	Suggested Use
1	SQL	Adjusts squelch level (filters static)	Set between 2-4 to avoid noise
2	TXP	Transmit power: HIGH or LOW	LOW for close range, HIGH for distance
4	VOX	Enables voice-activated transmission	Hands-free use (bike/hike)
5	WN	Wide/Narrow signal width	WIDE = stronger, NARROW = cleaner
7	TDR	Dual watch mode (monitor 2 channels)	ON to listen to 2 channels at once
13	T-CTCS	Transmit tone for repeaters	Needed if repeater requires a tone
25	SFT-D	Repeater shift direction	Set to + or - based on repeater info
26	OFFSET	Offset amount for repeater access	VHF = 0.600, UHF = 5.000
30	SC-ADD	Add/remove channel from scan list	Use to skip inactive channels

Appendix C: Beginner's Radio Go-Bag Checklist

Here's a lightweight go-bag setup to keep your Baofeng ready for daily carry or emergencies:

Essentials
- Fully charged Baofeng radio
- Spare battery or high-capacity battery
- USB charging cable or car adapter
- Extra antenna (e.g., Nagoya 771)
- Channel reference list or laminated cheat sheet
- Small notebook and pen for logging contacts

Optional Add-Ons
- Earpiece or speaker mic
- Compact flashlight or headlamp
- Backup power bank (USB)
- Mini first-aid kit
- Lightweight shoulder pouch or carrying case

Whether you're hiking, volunteering, or riding out a storm, this kit ensures you'll be ready to communicate instantly.

Appendix D: Practice Scripts for Talking On the Air

One of the biggest things that holds new radio users back is wondering, "What do I say?"

The truth is, you don't need fancy lingo. You just need a few clear, simple phrases to get started. Most hams appreciate plain speech that's polite, respectful, and easy to understand.

These scripts are here to help you practice, or use them as-is when you're on the air.

Basic Calling Script

Use this to see if anyone's listening.

"This is [your callsign], monitoring [frequency or repeater name]."

Example: "This is KJ7ABC, monitoring 146.520."

You can say this once, wait 10–15 seconds, then repeat once more if no one answers.

Radio Check Script

Use this to test your radio and confirm someone can hear you.

"This is [your callsign], radio check please."

If someone responds, they might say: *"KJ7ABC, you're coming in loud and clear."*, or *"KJ7ABC, I hear you but you're a little scratchy."*

Just thank them and sign off or continue the conversation.

Responding to a Call

When you hear someone call out and want to reply:

"[Their callsign], this is [your callsign]. I copy you. How's it going?"

Keep it short and steady. You're just making a friendly connection.

Checking Into a Net

Many nets will guide you with exact wording, but here's a simple version:
"This is [your callsign], checking in. No traffic."

That means you're present but don't have anything to report. If you do want to chat or share a message, just say:

"This is [your callsign], checking in with traffic."

Then wait for net control to ask you to speak.

Signing Off

If you're done talking, let people know:

"Thanks for the contact. I'll be monitoring, but I'm signing off for now. This is [your callsign], clear."

That signals you're finished and leaves the channel open for others.

Practicing Off-Air

If you're still unlicensed or just nervous, practice the scripts aloud at home. You can even rehearse with a friend using FRS or GMRS channels (legally) to get comfortable.

The goal isn't to sound fancy. It's to be clear, calm, and confident. These scripts give you a starting point, and after a few conversations, you won't need them. You'll just speak like yourself.

Glossary

Analog Radio: A type of radio that sends your voice as smooth waves. Most Baofengs use analog signals.

Antenna: The 'stick' on your radio that sends and receives signals. It's how your voice travels through the air.

B

BCL (Busy Channel Lockout): Prevents you from transmitting if the channel is already in use.

Backup Battery: An extra battery you keep charged in case your main one runs out.

Band: A group of frequencies. Your Baofeng usually works on VHF and UHF bands.

Base Station: A radio setup that stays in one place, like at home or a command center.

Battery Eliminator: Lets you power your radio from a car's power outlet instead of using the battery.

Battery Save Mode: A setting that helps your battery last longer by using less power when the radio is idle.

Belt Clip: A clip that attaches to your radio so you can wear it on your belt.

C

CHIRP: A free computer program that helps you quickly program your Baofeng radio.

CTCSS: A private tone that blocks out other conversations. Like a password for your channel.

Call Sign: A unique ID you use on the air, usually assigned when you get a ham license.

Carrier: The invisible wave that carries your voice when you press the PTT button.

Channel: A specific frequency or memory slot where communication happens.

Charger: The dock or cable you use to recharge your radio battery.

Contact Log: A page where you can record who you talked to and when.

D

Decibel (dB): A way to measure signal strength or sound. More dB means stronger signal.

Display Mode: Switch between showing channels (like CH-001) or frequencies (like 146.520).

Dual Watch: Lets you listen to two channels at once.

Duplex: Transmits and receives on two different frequencies. Used with repeaters.

E

Emergency Alert Button: Sends an emergency signal when pressed.

Emergency Channel: A saved frequency used only in emergencies.

F

Firmware: The software inside your radio that controls how it works.

Frequency: A number that shows where you're talking or listening. Example: 146.520 MHz.

Frequency Step: The amount your radio jumps when scanning or tuning.

G

Grid Square: A map system used by hams to show location.

H

Handheld Radio: A portable radio like your Baofeng. Easy to carry and use on the go.

I

Interference: Noise or signals that mess with your transmission.

K

Keypad Lock: A setting that prevents you from pressing buttons by accident.

Keypad Tone: The beep sound you hear when you press buttons.

L

LCD Screen: The part of your radio that shows channels, settings, and battery info.

Legal ID: You must say your call sign regularly when transmitting, per FCC rules.

License (Ham): Permission from the FCC to legally transmit on amateur radio frequencies.

Line of Sight: How far your signal can go without being blocked by buildings or hills.

M

Memory Channel: A saved frequency you can quickly return to.

Mic Gain: Controls how sensitive your microphone is.

Monitor Button: Temporarily turns off squelch so you can hear everything on a channel.

N

Net (Radio Net): An organized radio check-in or group chat over the air.

O

Offset: The difference between transmit and receive frequencies for repeaters.

P

PTT (Push-to-Talk): The button you press to speak into your radio.

Programming: Saving frequencies and settings into your radio.

Programming Cable: A cable that connects your radio to a computer for easier programming.

Propagation: How radio waves travel, depending on weather, terrain, or time of day.

Q

Q-Signals (QRZ, QTH, QSL): Quick codes hams use. QRZ = 'Who's calling?', QTH = 'Location', QSL = 'Got it'.

R

RF (Radio Frequency): The range of frequencies your radio uses.

Radio Check: A test message to ask if your signal is coming through.

Radio Net: A regular on-air meeting or chat for ham operators.

Repeater: A device that rebroadcasts your signal farther.

Repeater Offset: The amount a repeater changes the frequency between talking and listening.

Roger Beep: A beep sound at the end of your transmission to say 'I'm done talking.'

Rubber Duck (Antenna): The short flexible antenna that comes with your radio.

S

Scan Function: Automatically checks channels to find one with activity.

Scan Resume: Tells the radio what to do after it hears a signal—pause or keep scanning.

Scrambler: A feature that makes your voice harder to understand without the same radio settings.

Signal Report: A way to describe how strong and clear a signal is.

Signal-to-Noise Ratio: The balance between your voice and background noise.

Silent Monitoring: Listening without transmitting. Great for learning or emergencies.

Simplex: Talking and listening on the same frequency.

Skip: When signals bounce off the atmosphere and travel much farther than usual.

Spurious Emissions: Unwanted signals your radio might accidentally send. New radios minimize this.

Squelch: Blocks background noise when no one is talking.

Standby Mode: Your radio is on and listening, but not transmitting.

Standing Wave Ratio (SWR): Shows how well your antenna is matched to your radio. Lower is better.

T

Timeout Timer (TOT): Limits how long you can transmit at once to avoid hogging the airwaves.

Tone: A code used to open repeaters or talk privately.

Transmit: Sending your voice over the radio waves.

U

UHF (Ultra High Frequency): Radio band good for cities and buildings. Short-range but strong.

USB Cable: A wire used to connect your radio to a computer.

V

VFO Mode: Lets you manually type in frequencies instead of using pre-saved ones.

VHF (Very High Frequency): Radio band good for open areas and long distances.

Voice Inversion Scrambler: A basic way to make your voice sound scrambled for privacy.

Voice Prompt: A voice that helps guide you through menu options on your Baofeng.

Volume Control: The knob that makes the sound louder or softer.

W

Weather Channel: A channel where you can listen to weather alerts and updates.

Whip Antenna: A long flexible antenna that improves range compared to the stock one.

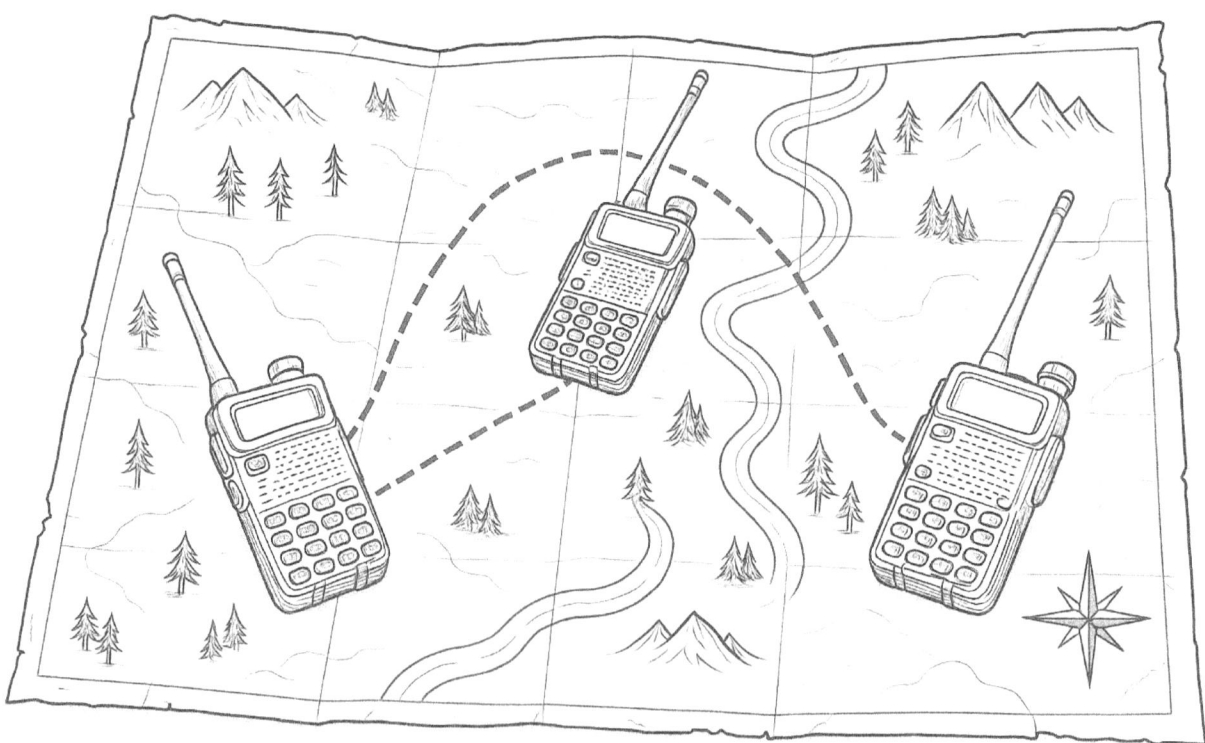

Baofeng Radio Contact Log

Your personal record of calls, contacts, and conversations.

Date	Time	Frequency	Callsign	Signal Report	Location	Notes

Baofeng Radio Contact Log

Your personal record of calls, contacts, and conversations.

Date	Time	Frequency	Callsign	Signal Report	Location	Notes

Baofeng Radio Contact Log

Your personal record of calls, contacts, and conversations.

Date	Time	Frequency	Callsign	Signal Report	Location	Notes

Baofeng Radio Contact Log

Your personal record of calls, contacts, and conversations.

Date	Time	Frequency	Callsign	Signal Report	Location	Notes

Baofeng Radio Contact Log

Your personal record of calls, contacts, and conversations.

Date	Time	Frequency	Callsign	Signal Report	Location	Notes

Baofeng Radio Contact Log

Your personal record of calls, contacts, and conversations.

Date	Time	Frequency	Callsign	Signal Report	Location	Notes

Baofeng Frequency Planner

Plan your channels for hikes, events, or group comms.

Channel #	Group Name	Frequency	CTCSS/DCS	Power	Notes

Baofeng Frequency Planner

Plan your channels for hikes, events, or group comms.

Channel #	Group Name	Frequency	CTCSS/DCS	Power	Notes

Baofeng Net Check-In Log

Track your weekly nets, repeaters, and signal reports.

Date	Net Name	Frequency	Repeater	Net Control	Signal Report	Notes

Baofeng Net Check-In Log

Track your weekly nets, repeaters, and signal reports.

Date	Net Name	Frequency	Repeater	Net Control	Signal Report	Notes

Baofeng Net Check-In Log

Track your weekly nets, repeaters, and signal reports.

Date	Net Name	Frequency	Repeater	Net Control	Signal Report	Notes

Baofeng Net Check-In Log

Track your weekly nets, repeaters, and signal reports.

Date	Net Name	Frequency	Repeater	Net Control	Signal Report	Notes

Baofeng Net Check-In Log

Track your weekly nets, repeaters, and signal reports.

Date	Net Name	Frequency	Repeater	Net Control	Signal Report	Notes

Baofeng Net Check-In Log

Track your weekly nets, repeaters, and signal reports.

Date	Net Name	Frequency	Repeater	Net Control	Signal Report	Notes

Baofeng Emergency Readiness Checklist

- ☐ Radio fully charged (main battery)
- ☐ Spare or high-capacity battery packed
- ☐ Antenna attached and tested
- ☐ NOAA weather station saved
- ☐ Local repeater frequencies programmed
- ☐ Earpiece or speaker mic packed
- ☐ Printed quick reference or net info
- ☐ Programming cable packed (if needed)
- ☐ Emergency channels saved and labeled
- ☐ CHIRP backup file saved to USB/cloud
- ☐ USB car charger or power bank included
- ☐ Compass, flashlight, or map packed
- ☐ Pen and paper or notebook for notes
- ☐ Emergency contact numbers written down
- ☐ Stored in a weatherproof case or pouch

Enjoyed This Book?

If this book helped you feel more confident with your Baofeng, I'd love to hear about it.

Leaving a quick review will help radio users who were in the same position as you when you started reading this book.

And the best part?

It only takes about 25 seconds.

One sentence about what you liked most and one sentence about your experience with your Baofeng radio.

Your feedback helps other beginners find a guide that actually makes sense

Thanks again for being here. You're not just learning radio… you're becoming someone who can step up when it counts.

www.ingramcontent.com/pod-product-compliance
Lightning Source LLC
Chambersburg PA
CBHW062314220526
45479CB00004B/1159